Roger Cotes – natural philosopher

Fig. 1. Bust of Roger Cotes, in the Wren Library at Trinity College, Cambridge. (By permission of Trinity College Library, Cambridge.)

Roger Cotes – natural philosopher

RONALD GOWING

CAMBRIDGE UNIVERSITY PRESS

Cambridge
London   New York   New Rochelle
Melbourne   Sydney

Published by the Press Syndicate of the University of Cambridge
The Pitt Building, Trumpington Street, Cambridge CB2 1RP
32 East 57th Street, New York, NY 10022, USA
296 Beaconsfield Parade, Middle Park, Melbourne 3206, Australia

First published 1983

Printed in Great Britain at the University Press, Cambridge

Library of Congress catalogue card number: 82-1154

British Library cataloguing in publication data

Gowing, Ronald
Roger Cotes.
1. Cotes, Roger   2. Mathematicians—Great
Britain—Biography
I. Title
510′.92′4   QA29.C
ISBN 0 521 23741 6

As a young man, Cotes had that density of thought that occurs only rarely and is the hallmark of the genius who departs before his time.

(J. E. Hofmann, 'Weiterbildung der Logarithmischen Reihe Mercators in England, III', *Deutsche Mathematik*, 5 (1940) 368–75.)

Contents

Preface

My interest in the work of Roger Cotes arose during a course of lectures in the History of Mathematics, given by Dr J. M. Dubbey at University College, London. That Sir Isaac Newton, in his late sixties, acknowledged as among the greatest mathematicians of his day, President of the Royal Society, Master of the Mint, should agree to the appointment of Roger Cotes, twenty-seven years old and relatively unknown, as editor of the long awaited second edition of *Philosophiae Naturalis Principia Mathematica*, was sufficiently intriguing to prompt a closer look at Cotes' achievement. It is my hope that the present volume will make some contribution to an accurate assessment of this.

In preparing the book I received help and encouragement from a number of scholars in the increasingly important field of the history of mathematics. In particular, I wish to express my warmest thanks for the help, advice and personal kindness of Dr D. T. Whiteside, who has created, in his monumental *The Mathematical Papers of Isaac Newton* (Cambridge University Press, 1967–80, 8 volumes), a rich source to which future scholars will, as I do, return again and again.

Such skill in Latin as I possess is that of the enthusiastic but self-taught amateur. However, the translation of Logometria, which I offer in Appendix 1 owes much to the professional checking and correcting of Mrs Stephana Babbage, of Cambridge, and I gratefully acknowledge this help, laying claim only to any surviving errors.

Dr Gaskell and his staff at the Wren Library of Trinity College, Cambridge, have been ever courteous, patient, and most helpful. I offer them my sincere thanks for this, and for permission to publish some illustrative material. I am pleased to extend these thanks and acknowledgements to the Librarian of Clare College, Cambridge, and to the staffs of the libraries of the Universities of Cambridge and London, the Royal Society, the Royal Greenwich Observatory, and the British Library.

To the staff of Cambridge University Press, I would like to offer my especial thanks for their interest, care and great patience, which have made it possible for this book to match the high standards we have come to expect from the Press.

R Gowing
Bickley
1981

Introduction

Roger Cotes, one of that impressive band of able and distinguished men and women who have emerged from English rectories, was a Fellow of Trinity College, Cambridge, at the age of twenty-five, and the first Plumian Professor of Astronomy and Experimental Philosophy at the age of twenty-six. He is best known for his meticulous and creative editing of the second edition of Newton's *Principia* (Cambridge, 1713) and for his notable Preface to that work. Contemporary opinion places him among the most able of British mathematicians of his day, and close study of his work supports that view. His influence on the subsequent development of mathematics was much less than his ability and achievement merit: his name survives in 'the Cotes property of the circle', 'the Cotes spiral', 'the Cotes–Newton formulae for approximate quadrature'; otherwise, the man and his work are relatively little known. It is the aim of this book, in this the tercentenary year of his birth, to make his work more widely known, and so to reveal something of the man.

There is little biographical information. The good article by Agnes Mary Clarke, in the *Dictionary of National Biography*, refers to most of what is available and is largely, but not entirely, correct. We do, however, have the remarkable, fascinating and almost complete correspondence between Cotes and Newton, written during the four years in which Cotes was editing *Principia*, from Cotes' first tentative letter in 1709, to the final rather chilly exchange about the errors after the publication of the second edition in 1713. The correspondence is preserved in the library of Trinity College, Cambridge, and in the Cambridge University Library. It has been superbly edited and published in *Correspondence of Sir Isaac Newton and Professor Cotes*, by J. Edleston, published by Frank Cass and Company in 1850 (and in a new impression as number 12 of the Cass Library of Science Classics, in 1969). Excellent and detailed commentaries have been provided by I. B. Cohen in *Introduction to Newton's Principia* (Cambridge, 1971), and by Rupert Hall and Laura Tilling in *Correspondence of Sir Isaac Newton*, vol. 5 (7 vols., Cambridge, 1975). It would be presumptuous of me to attempt to recapitulate or add to these distinguished works. In Chapter 1 I have drawn together such parts of the correspondence as throw light on Cotes' personality, the circumstances surrounding his appointment as Plumian Professor, and his relationship with Newton. It was against the background here sketched that Cotes was preparing his own work for publication, and the one paper he submitted for publication in his lifetime, Logometria, is discussed in Chapter 2. The original Latin paper is fairly readily available in *Harmonia Mensurarum, Sive Analysis & Synthesis per Rationum & Angulorum Mensurae Promotae*, ed. R. Smith (Cambridge, 1722), pp. 4–41, but an English translation has not previously been published; the full translation is therefore given in Appendix 1. In Logometria, Cotes developed the theory of logarithms, following earlier work, particularly that of Halley and De Moivre. J. E. Hofmann says in his article 'Weiterbildung der Logarithmischen Reihe Mercators in England, III', *Deutsche Mathematik*, 5 (1940–1), 368–75: 'As a young man Cotes had that density of thought that occurs only rarely and is the hallmark of the genius who departs before his time.' The principal application of the theory was to the integration of certain rational and irrational algebraic forms, yielding logarithmic, trigonometric and hyperbolic functions, functions for the most part not then clearly recognised. The power of the new methods was demonstrated in the long Scholium Generale of Logometria, where the solutions of a number of problems of contemporary interest were given, without proof. The proofs depend upon the eighteen tables of integrals developed by Cotes. These integrals and their application form the subject of Chapter 3; the eighteen fluxional forms which Cotes integrated are shown in Appendix 3.

The tables of integrals were published posthumously in 1722, by Robert Smith, Cotes' cousin, assistant and eventual successor to the Plumian chair, in *Harmonia Mensurarum*. In this work, Smith collected together most of Cotes' surviving papers; I have translated and worked through the whole of this work. The first three parts of the book are often referred to as Logometria, Parts one, two and three. Part one is Logometria almost exactly as it appeared in the *Philosophical Transactions of the Royal Society*, vol. 29, no. 338 (1714), pp. 5–45. Part two consists of Cotes' eighteen tables of integrals and Part three is a collection of twelve illustrative problems in which the application of the integrals to the solutions is shown in detail. All this is discussed in Chapter 3. Cotes was busy preparing his papers for publication at the time of his death in 1716. His discovery of a general method of factorising $a^n \pm x^n$ into linear and quadratic factors for integral values of n, enabled him, towards the end of his life, to extend his work on integrals. Smith edited and further developed this later work, arranging the results in ninety-three tables, with a good deal of needless elaboration. I refer to the forms here integrated as the Cotes–Smith Forms: they form Part four of *Harmonia Mensurarum*. The tables are discussed, together with the theorem on which they depend, in Chapter 4.

As Professor of Astronomy, Cotes carried a large share of the responsibility for the construction and equipping of the observatory which was constructed on top of the King's gate at Trinity College. He seems to have entered upon this work with great enthusiasm, but there is little record of practical work carried out. Such as there is I have gathered together in Chapter 5, in which I have also discussed the tract, Aestimatio Errorum in Mixta Mathesi, published in *Harmonia Mensurarum*. This last is a serious attempt to apply fluxional concepts to a study of the errors arising in astronomical observations. Cotes' work on *Principia* led him to consider ways of computing cometary orbits from a limited number of observations, and to an awareness of the need for more accurate lunar tables. Agnes Mary Clarke states in the *Dictionary of National Biography* that Cotes remodelled Cassini's tables, and had undertaken to prepare new tables of the moon. I have found no record of this work, but Cotes did write two tracts concerned with the construction of tables. These are De Methodo Differentiali Newtoniana, and Canonotechnia (or, The Construction of Tables by Differences). Both these papers are printed in *Harmonia Mensurarum*, and formed part of the public lectures which Cotes was required to give by the statutes of his appointment. The first was largely completed before the publication of Newton's tract on the *Method of Differences* (London, 1711) and, having seen Newton's paper, Cotes added the section which contains the so-called Cotes–Newton formulae for approximate quadrature (Simpson's rule, the 'three-eighths rule', and so on). Canonotechnia is one of the most interesting of Cotes' papers, and deals with more general methods of interpolation and sub-tabulation. Both are discussed under the heading Numerical methods, in Chapter 6.

Besides the papers appearing in *Harmonia Mensurarum*, there is one other published work by Cotes, namely, his *Hydrostatical and Pneumatical Lectures*. These were published in 1738 by Robert Smith (see second edition, Cambridge, 1747), as he says in his Preface, to forestall a

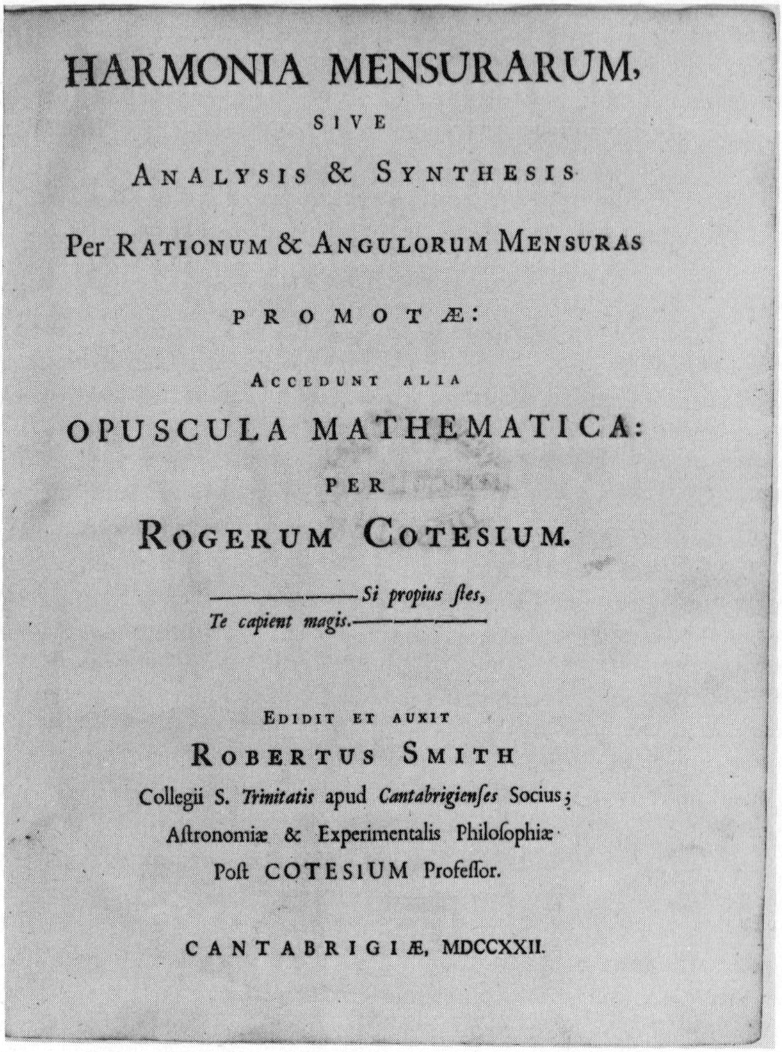

HARMONIA MENSURARUM,

S I V E

ANALYSIS & SYNTHESIS

Per RATIONUM & ANGULORUM MENSURAS

P R O M O T Æ:

ACCEDUNT ALIA

OPUSCULA MATHEMATICA:

P E R

ROGERUM COTESIUM.

—————————— *Si propius ſtes,*
Te capient magis.—————

EDIDIT ET AUXIT

ROBERTUS SMITH

Collegii S. *Trinitatis* apud *Cantabrigienſes* Socius;
Aſtronomiæ & Experimentalis Philoſophiæ
Poſt COTESIUM Profeſſor.

C A N T A B R I G I Æ, MDCCXXII.

Fig. 2. Title page of *Harmonia Mensurarum*, Cambridge University Press, 1722.

pirated edition. The lectures were intended for undergraduates and are at a fairly elementary level, although full of interesting detail. My brief comments about them, together with some more general remarks, fit most readily into the final chapter.

1

Cotes as first Plumian Professor, and as editor of Newton's Principia

The posthumous publication of Christian Huygens' *Cosmotheoros* (The Hague, 1698) had an unexpected consequence in influencing Dr Thomas Plume, the then Archdeacon of Rochester, to bequeath nearly £2000 to found the Plumian Professorship of Astronomy and Experimental Philosophy, at Cambridge. The first and, no doubt, all subsequent holders of the chair, were surely intrigued by Huygens'

It is hardly possible that an adherent of Copernicus should not at times imagine that it seems not unreasonable to admit that, like our globe, the other planets are also not devoid of vegetation and ornament, nor perhaps of inhabitants. It is not necessary to assume that the equipment of other planets is fundamentally different from what we know on earth . . . If there are intelligent beings the rules of geometry must be the same as for us.

(C. Huygens, *Cosmotheoros* (The Hague, 1698), Preface.)

For an Archdeacon of Rochester, Doctor of Divinity, to found a chair of Astronomy and Experimental Philosophy, possibly in the hope of resolving such questions, is a nice comment on the pragmatic and empirical nature of some English philosophy at the time.

The young and able mathematician appointed to be the first Plumian Professor was Roger Cotes, Fellow of Trinity College at the age of twenty-five (in 1707), admirer and follower of Newton, protégé of Richard Bentley (the colourful, controversial and combative Master of Trinity), friend of William Whiston and first Plumian Professor at the age of twenty-six.

Roger Cotes was born 10 July 1682, the second son of Robert Cotes, rector of Burbage in Leicestershire, by his second wife Mary, formerly Mary Paget, née Chambers. Because of his mathematical ability, as a boy of eleven or twelve years, Cotes was moved from Leicester School and sent to be coached by his uncle, John Smith, rector of Gate-Burton in Lincolnshire. John Smith had married Hannah Cotes, Roger's aunt, at about the time of Roger's birth. It must have been a very congenial household. There was a girl Elizmar, about Roger's age, and a boy Robert, some seven years younger. This trio were all to have their lives closely interwoven with Trinity College, Roger and Robert as first and second Plumian Professors, Robert also as a very distinguished Master of the College and founder of the Smith's Prize; Elizmar was to be Robert's housekeeper in the Master's Lodge for many years. All three now lie buried in the College chapel or antechapel.

John Smith was an able and kindly tutor, and his influence on Cotes must have been considerable. At the age of sixteen, Cotes, then at St. Paul's School in London, wrote authoritatively and knowledgeably to his uncle about mathematics books then available. Cotes and his uncle clearly share a common enthusiasm, and the tone of the letter is of one mathematician to another. Smith had evidently asked Cotes to look out for works by Galilei and Kepler. In his reply, postmarked 31 December 1698 ([4], Letter xcv), Cotes explains that these works are 'dispersed in divers Volumes, put forth at different times', and includes with his letter a long list of such works culled from catalogues then available: 'You may from hence pitch upon those you most like of, & I shall be very glad to use my utmost endeavours to procure 'em for you.' He himself has Galilei's *Nuncius Siderius* (Venice, 1610), together with Kepler's *Dioptricks* (Augsberg, 1711) and Gassendis's *Astronomy* (Paris, 1647): 'If you please, I will send you 'em.' He has also Wallis's *Algebra* (3 vols., London, 1685), with which he is very pleased. After a brief account of the book, he adds, with all the authority of his sixteen years, and that authority appears already to be considerable: 'In my mind there are many pretty things in that book worth looking into'. The letter is interesting for the indication it gives of Cotes' reading at the threshold of his career, but it is also interesting from another point of view. It marks the start of an enquiry which was to absorb a great deal of Cotes' attention during the next few years; the development of methods of quadrature, or squaring of curves, i.e., of integration. John Smith, able coach, had asked the right question: 'You wrote of ye Quadrature of Curve's, as yet I cannot enquire of any Mathematician about 'em.' But Cotes, already showing that thoroughness which was to be such a distinguishing mark of his work, did not leave it there: 'Sr Edw: Sherbourne in his Appendix to his Translation of Manilius's *Astronom*: tells us that from Mr Isaac Newton is expected a New general Analytical Method by infinite Series for ye Quadrature of Curvilinear figures.' And in Wallis's Algebra also: 'Amongst other things he speaks of squaring Curves.'

Four years later, 6 April 1699, Cotes entered Trinity College, Cambridge, as a pensioner and, during the long vacation between his second and third years, wrote to his uncle giving a very able exposition of: 'The Method of Fluxions applied to the buisness of Quadratures'. Cotes is enthusiastic and exuberant:

This may p'haps serve as a specimen of ye Method of Fluxions applied to the buisness of Quadratures tho its uses seem to be as inexhaustible as they are Naturall & Easy for by it the great Geometers of our Age are enabled To draw Tangents, To rectifie, To find ye Evolutes, The Caustics by reflection and refraction of all sorts of Curves, To measure ye Surfaces generated by their rotation. The solids they comprehend. The Centers of Gravity, Oscillation & Percusn. of all these. To resolve all sorts of Questions de Max & Min. To find the Points of Inflection & Rebroussement (as y^e French term it) in all Curves & y^e Converse of all these & many more. But what wonders does it not do when applied to Nature! where it triumphs alone & admits of no Partner – But I transgress y^e Bounds of a Lett^r Pray S^r pay my humble respects to my Aunt; and my Love to Cozz^{ss}.

$$Y^r \text{ very \&c}$$
$$R. \text{ Cotes.}$$

This to the Reverend Mr Smith Rector
of Gate-Burton near Gainsborough
by Newark, Caxton.
(Cotes to John Smith, 9 September 1701 [4], Letter XCVII.)

It is not surprising that in an earlier letter, John Smith had been gently ironic: 'we ... count it as a favour y^t you can spare us any share of your affection from your dear Mrs Mathesis; I am glad to hear y^t she so easily yields to Your courtship, and has procured You such signal marks of favour from great men as Dr Bently Mr Hanbury' (John Smith to Cotes, 30 August 1701 [4], Letter XCVI).

The approbation of Richard Bentley was a significant influence on Cotes' career, as was that of William Whiston. Nathaniel Hanbury seems to have been a relatively minor figure in this respect, although John Smith reckoned him a better mathematician than Whiston, and regretted that he had not been appointed 'vice-professor' in Newton's place. Smith, in his Lincolnshire village, seems to have kept himself surprisingly well informed. Whiston had in fact been appointed Newton's deputy and begun his duties, which required him to lecture in Newton's stead, early in 1701 during Cotes' first year. Newton resigned his professorship, and his fellowship, later in the same year. Whiston was appointed Lucasian Professor.

Cotes, with his young enthusiasm, responded warmly to Whiston. Whiston ('paradoxical to the verge of craziness, but intolerant to the verge of bigotry'), publicist, polemicist, and populariser of 'Newtonian philosophy' was a dominant figure, differing fundamentally in temperament from the more gentle, modest and, at times, tentative Cotes. Whiston formed a high opinion of Cotes' mathematical ability, later describing himself as 'but a child to Mr. Cotes', and Cotes found in Whiston a certainty and ardour which led and encouraged him in his own complete acceptance of the Newtonian philosophy. The two men became close friends. Together they wrote a course of lectures in mechanics, i.e., hydrostatics and pneumatics. Whiston described Cotes' contribution as far superior to his own, and the lectures are notable for being part of one of the earliest courses in which students were expected to carry out practical work and experiments (see Chapter 7). Cotes, scholarly and enthusiastic, Whiston, flamboyant, extrovert and competent, would have made an impressive team. When, in later years, Whiston was in dispute with the College and the University authorities because of his religious beliefs, Cotes urged him (without effect) to take a more moderate attitude. After Whiston had resigned his chair, which he did in 1710 as a result of the religious controversies, the friendship continued. In correspondence with Whiston, Cotes gave firm and assured advice on a number of matters, in particular those connected with the problem of determining longitude, and with some aspects of Whiston's public lectures. The brief biographical notes on Cotes in *The General Dictionary, A New and General Mathematical Dictionary*, eds. W. Whiston, T. Osborne and others (London, 1761), were probably written by Whiston.

The appointment, on 1 February 1700, of Richard Bentley, eminent classical scholar and King's Librarian, to be Master of Trinity, was of great significance for Cotes personally. Bentley's aim, to raise the academic standard of the work of the College, and to establish it as a centre for all kinds of learning and, in particular, of Newtonian philosophy, was pursued with all of Bentley's immense energy, imagination and skill. In part of his project to encourage the study of the physical sciences in general at Trinity, he was successful in three major undertakings concerning Cotes: the appointment of Cotes (by then a Fellow of Trinity) as first Plumian Professor; the construction of the observatory, under Plume's bequest, at Trinity; and the appointment of Cotes as editor of the much awaited second edition of Newton's *Principia* (Cambridge, 1713).

In his will, Dr Plume had stated:

All this money I give and bequeath to erect an Observatory and to maintain a studious and learned Professor of Astronomy & Experimental Philosophy, and to buy him and his successors books and Instruments, Quadrants, Telescopes &c and to buy or rent or build an house with or near the said

Observatory, which Observatory is to preserve the said Instruments in from time to time for the use of the said Professor, for the time being, and the house to reside in or if he be resident in a Colledge, to let and to receive the rent for himself, but so as any ingenious Scholars or Gentlemen may resort to him in the proper seasons to be instructed and improved by him in the knowledge of Astronomy, the Globes, Navigation, Naturall Philosophy Dialling & other practical parts of the Mathematicks in or near Cambridge. And that he be obliged to read in the Latin tongue, one lecture every term and at least two in every year, making or proclaiming a suitable Oration in Latin before each lecture to recommend the said Sciences to the study of his Auditors & to print or leave a fayre manuscript copy in quarto, one copy every year, in the University library to be preserved there amongst their archives and to [be] bound up together when they shall come to a fit Volume, by himself or his successors as they were delivered in the Physick Schools or elsewhere according to the appointment of the Vice Chancellor. of the Reading of which lectures the Beadles to give notice in every Colledge as they do for Concords or Plenum. for which the Professor shall content them, or else shall cause onely the Schols bell to be rung or both as the Vice Chancellor shall order. And I will y^t Dr Francis Thompson, Dr Covell Master of Christs Colledge and his successors there, The Mathematicall Professor for the time being, Dr Bentley Master of Trin Colledge with the advice of Mr Newton in London, Mr John Ellis of Caius Colledge, and Mr Flamsteed the Royall Mathematician at East Greenwich, or their successors shall consult with or about such Statutes and Orders for the Election, Residence qualifications & performances of the said Professor as shall be most requisite for the perpetuation and benefit of the said Professor and the Improvement of Astronomy and Naturall Philosophy enjoyning him to keep a boy or two to assist him in his observations & to teach him Physicks Mathematicks Navigation and Astronomy: If I settle not these statutes myself I will that the Principall mony be not removed from the Bank till they have a very good purchase to make with it. This is my last will and testament.

Signed in the presence of

William Daws   Octob: 20: 1704.
Francis Gillot
Tho Lomax

(Flamsteed Correspondence, Royal Greenwich Observatory, vol. 15, fo. 158.)

(Plume died November 1704.)

John Flamsteed, the Astronomer Royal, engagingly referred to by Plume as 'The Royal Mathematician', and indeed so signing himself at the end of the following letter to Whiston, had expected that matters relating to the founding of the Professorship, and in Flamsteed's view this certainly included electing the first Professor, would be settled by himself and Newton, acting as trustees. He was irked to find himself only one of seven, and that in an advisory role: the situation was unclear. Were all seven to be trustees? Were all seven to be electors? Doubts about these matters gave rise to a certain amount of acrimony, not lessened by Bentley's manoeuvring to have the Professorship and the observatory established at his own college. Flamsteed felt at a disadvantage. It was clear that the real power would lie with the Masters of Colleges for whom the opinion of the Lucasian Professor, (Whiston) and of Newton would be crucial. Flamsteed, an irascible and difficult man, had enjoyed (the verb is suitable) strained relations with Newton for some years. One of the central issues was the relative importance each attached to observational, as opposed to theoretical, astronomy. Newton needed Flamsteed's observations to develop further his work on the theory of the moon's motion, and suggested that this would bring fame and honour to Flamsteed for the accuracy of his observations. Flamsteed's view was the reverse – the accuracy of his observations would be the test of Newton's theory. This is a gross over-simplification of the dispute, but it rumbled on for many years, flaring to a climax when Flamsteed, dissatisfied with the eventual editing by Halley of his observations in *Historia Coelestes* (see J. Flamsteed *Historia Coelestes* (2 vols., London, 1712–23)), separated out the offending parts from some three hundred copies of the book, and burned them in 'a bonfire to heavenly truth' (see F. Baily, *Account of the Rev. John Flamsteed* (London 1835)).

In the matter of the appointment of the Plumian Professor, Flamsteed feared (with some justification) a *fait accompli* on the part of the Masters of the Colleges. He was anxious to see the appointment of his own assistant, John Witty, who had, it must be acknowledged, then almost unrivalled opportunities for practical astronomy. Whiston was obviously the best man to approach, which Flamsteed did in the following letter:

Flamsteed to Whiston

> The Observatory
> Feb: 13: 1705/6

Sir

I have been told lately that the Masters of Trinity Kays and Christ have proposed an ingenious young man of Trinity for the Astronomy professorship given by my late Friend and neighbour Dr Plume, who discoursed sometimes with me of this his design; and fixed upon an ingenious Young Man that about 3 years ago was my servant, for his professor. But upon my informing him that the said person, tho' otherwise very proper for it, had then but a Moderate knowledge of the Latin Tongue, and none of the Greek, laid by his thoughts of him; and told me he had left Sr Isaac Newton and my self Trustees to see his will executed in this particular. I find now he has added the Masters of the Colleges abovementioned with his old Friend Dr Tomson and your self. I hope the Masters in the University design not to determine any thing concerning his professor without consulting those who are equally concerned with them in the will of Dr Plume and more particularly informed of his design, and that they have not so far ingag'd themselves to any person but that there is still room to admitt one better qualified than that person proposed, if such an one offer himself. I esteem you a person of so much candour and sincerity, that I persuade myself you will wave all particular intretes, to serve the generous design of the Revd Donor of the professorship, and therefore send this to acquaint you that Mr John Witty formerly of St Johns College in Cambridge, has lived with me about 12 months: he left your University about 5 years ago, apply'd himself to the study of Geometry and Algebra in the Country, is a good proficient in 'em, is very ready at numbers, and since I entertained him has been continually employed in deriving the moon: and planets places from the observations here made in order to have 'em printed with my works, and other Astronomical Matters, and studys, whereby he has attain'd that knowledge which the D^r respected as absolutely necessary in his professor, and which probably is scarce to be attain'd elsewhere. On which account that I may answer the trust reposed in me by the deceas'd I thought myself obliged to give you this information concerning him, and to beg the favour of you to acquaint the trustees in Cambridge with it, and interpose with 'em that his merit may be considered with the other Candidates. I have been at London these 12 weeks by reason of frequent pains in my Head, and lat'ly in my feet, so I have not had an opportunity of proposing Mr Witty to Sr Isaac Newton; but if God spare me health I shall have occasion to wait upon him in a few days, and then I shall acquaint him with Mr Witty's desires to serve the University in this place, and his fitness for it above others of which I doubt not but he will be easily satisfied. in the meantime I intreat you not to think that this desire to prefer him proceeds from any interest I have to do it, but only from a sincere desire to answer the trust repos'd in me, and the rest of the trustees may do the same, on which account alone this comes to you from

Sr your affectionate and humble servant
John Flamsteed M R

To ye Revd Mr Whiston. professor of Mathematics in Cambridge.

(Flamsteed's Correspondence, Royal Greenwich Observatory, vol. 15, fo. 130.)

The cause of Flamsteed's concern was almost certainly the following agreement, preserved in the Cambridge University Library (Guard Book, C U R 39 12):

Wee the Master and Seniors of Trinity College Cambridge Doe Covenant and agree with the Trustees or Electors for the Professorship of Astronomy and Experimental Philosophy lately founded by Dr Plume, that Roger Cotes Fellow of this College now nominated to the said Professorship, and all his Successors after him of what College or Place soever they shall be, shall have the rooms and leads of the Kings Gate of this College, for a dwelling and observatory, soe long as the Trustees and Electors of the said Professorship shall think fit, the Professors paying from time to time ten poundes per annum to the Fellow whose Chamber it shall be in the College. Provided always, while they shall use the Kings Gate for the Observatory, that the Scholar appointed to be the Professors Assistant and to lodge in the same dwelling with him, be one of this College, to be chosen by the Professor with the consent of the Master. In testimony whereof we have affixed hereunto the College Seale the Nynth day of February Anno Dni 1705.

The date is (probably) Old Style, the year is 1706, and 9 February is about a week before Flamsteed's letter to Whiston, quoted above. Concern for John Witty was a lost cause, but Flamsteed did not give up. He recorded in his diary for 15 March the same year: 'Met Dr Bentley at Garraways: Sir Izaark Newton was there: we discoursed first about Dr Plumes Astronomical Professorship: the Doctor would have had my hand to a paper for the election of Mr Cotes: I refused till I saw him: he told me Mr Whiston and Mr Cotes should wait on me next week' (Flamsteed's Diary, Royal Greenwich Observatory).

Flamsteed's fellow 'adviser', Francis Thompson, was also active in advancing Flamsteed's views about correct interpretation of the will, and that to 'buy, build or rent a house for the Professor' meant just that. In a long apologia to Thompson, Sir John Ellis, Master of Caius tried to mollify Thompson by answering all his points, and eulogising the choice of the Trinity gatehouse, the selection of Cotes, and above all Bentley, to whose schemes the Cambridge group and Newton were clearly completely committed.

In a nice mixture of semantics and expedience, Ellis points out the many financial advantages of the scheme (see Flamsteed Correspondence, Royal Greenwich Observatory, vol. 69, fos. 196–8):

If an observatory be not according to the strictest sense of the word (Erected) nor a house for the Professor (Built) yet the Gatehouse at Trinity does fully answer and better the pious intention, and may very well deserve the name from the use to which it may now be appropriated. And yourself in viewing it, or hearing the description of it seemed much to approve the choice.

I presume you have not forgot all that great pains you took in walking about the town seeking a place for an observatory, and yet at last could find none that in every respect was comparable to what may be had now from this gatehouse, without such charge as would too much diminish the gift.

Putting up new buildings would reduce the money available for instruments, and reduce the income, making it difficult to recruit a suitable Professor, whereas the gatehouse offered spacious and convenient accommodation, and all for ten pounds per year: 'Well was it for the University that such an uncommon genius so active and fitted for great things as Dr Bentley became concerned . . . He is the main Wheel or Spring . . . and may be looked upon as the chief foster father of the designe.' Ellis' eulogy of the gatehouse continues: 'What honour doth it give to the study of Astronomy, and what honour to the founder of the Professorship when the best College (perhaps in the world) and of Royall foundation, parts with such a piece, and of such stately and magnificent structure, and suffers it to have the name of the Plumian Professor?' And, of more immediate interest to the present account:

Now are we liable to fuse a charge from pitching upon such a Professor. His years indeed are under 30, but his constant esteem and respect among his fellow collegians may evidence enough his good temper, and he is remarkable besides for learning sobriety, and indefatigable diligence. His inclination love and delight from a child hath been in the Mathematics in which he hath the reputation of being eminent above all his contemporaries throughout the whole University. Besides, he hath something of an Estate, to secure him from being pinched while carrying on costly Experiments.

That some amends may be made for this my scanty character, let me refer it to Mr Whiston, whose judgement is unquestionable and acquaintance with him uncomparably beyond mine.

And then, curiously: 'I hear not of any weakness in his constitution, portending that he may not last long under his profession. If he does we all may hope, that his fame spreading to foreign countrys as well as our own, may bee much honour to his founder for his generous gift.' If Flamsteed saw this long letter, he must have smiled wryly at the mention of Cotes' Estate, 'to keep him from being pinched whilst carrying out costly experiments'. Flamsteed had suffered from similar assumptions. Ellis' purple prose was no emollient. Even this passage:

I confess that more care should have been taken in consulting yourself and Mr Flamsteed, but I had heard that Sir Isaac, Dr Bentley and he had had a meeting [probably that on 15 March recorded in Flamsteed's diary, some nine months earlier] and conference about ordering the gift, and likewise that information had been given to yourself.

failed to pacify, and Thompson replied (see Flamsteed Correspondence, Royal Greenwich Observatory, vol. 33, fo. 74) with a condemnatory blast. Thompson commends, ironically, Ellis' zeal for the public good, and criticises him roundly for letting his judgement be 'swayed by a private friend' (Bentley? almost certainly). To quote dictionary definitions, when the meaning of words in the will was plain for all to see, was pedantry. The place selected for the observatory neither answered the manner nor the design (intention) of the will. The bequest was for the benefit of the whole University, and not for the private advantage of any one particular college. Mr Flamsteed had said that: 'The gatehouse was not fit for it, nor would save much money.' St John's gatehouse was preferable, the Virtutis gateway at Caius better than either. Christ's (where Dr Plume was) Peterhouse, Jesus, Emmanuel could all 'afford fair prospects', and many of the rest might also have good ideas for themselves. A separate building could be put up on a piece of ground sixteen or twenty yards square. More time should be taken: 'In that time also, a Petitioner may accomplish himself – with more skill and Experience, to become the first Professor.' But it was all sound and fury, signifying very little. Bentley's schemes matured inexorably. Cotes was elected first Plumian Professor (16 October 1707) and the observatory was built on the Trinity gatehouse; John Witty escaped into a decent obscurity.

From shortly after his appointment until within three years of his death, in fact from 1709 to 1713, much of Cotes' time and attention was taken up with the task of editing the second edition of Newton's *Principia*. In Bentley's project to establish Trinity College as a great centre of learning, particularly in the physical sciences, Cotes was Bentley's man. Bentley was prominent among those urging Newton to publish a second and revised edition and, after abandoning hopes of himself becoming the editor, succeeded in getting Newton to appoint Cotes. The devoted care which Cotes brought to the work more than justified Bentley's trust and, indeed, patronage. The second edition which resulted, described by Rupert Hall and Laura Tilling [5] as, 'to all intents and purposes, the *Principia* of subsequent history', was a fine achievement, and brought further renown to the College (and some financial gain to Bentley).

We have little record of the early discussions between Newton and Bentley concerning the plans for the second edition. Bentley quickly realised that he himself had neither the time nor the ability for the task, and saw in Cotes one who would be not only competent, but also amenable:

You need not be so shy of giving M^r. Cotes too much trouble: he has more esteem for you &c obligation to you, than to think y^t trouble too grievous, but however he does it at my Orders, to whom he owes more then y^t. And so pray you be easy as to y^t. We will take care that no little slip in a Calculation shall pass this fine Edition.

(Bentley to Newton, 20 October 1709 [5], p. 7.)

Two months earlier, Cotes, anxious to begin work, wrote to Newton, urging him to send the corrected copy for the printers, which had already been promised a month ago. Thus began that fascinating and illuminating correspondence between Newton and his editor, first published by Edleston in 1850 [4], and reissued in facsimile in 1969.

A very full discussion of the circumstances surrounding the publication of the second edition of *Principia* has been given by I. B. Cohen in [2], and the Cotes–Newton correspondence is discussed in detail by Rupert Hall and Laura Tilling in [5]. Cotes' drafts of his letters are preserved, together with Newton's replies, in the Trinity College Library MS R 16 38) and some additional correspondence is in Cambridge University Library (Add MS 3983).

In his first letter, Cotes, after urging Newton to send the copy, goes on firmly, if ingenuously, to say that he has worked through the forms of fluents in Newton's *De Quadratura Curvarum* (London, 1704) agrees with all except two, for which he gives the corrections. Newton replied mildly and courteously, thanked Cotes for the corrections, and said he had despatched the greater part of the copy for printing, via Mr Whiston, adding:

I would not have you be at the trouble of examining all the Demonstrations in the Principia. Its impossible to print the book w'out some faults & if you print by the copy sent you, correcting only such faults as appear in reading over the sheets to correct them as they are printed off, you will have labour more than is fit to give you.　　* * *

(Newton to Cotes, 11 October 1709 [4], p. 5.)

Cotes clearly had no intention of confining himself to the more or less clerical labour of 'printing by the copy sent him' and proofreading the sheets. The correspondence reveals him as conscientious, careful, able, thorough to the point of fussiness. Gently but firmly persistent, he argued with Newton, sometimes returning to the same point in successive letters, and frequently gaining the point. When he regarded the wording as unclear, he proposed changes; his patience and his admiration for Newton seemed boundless:

I never think the time lost when we stay the press for his further corrections & improvements of so very valuable a book, especially when this seems to be the last time he will concern himself with it.

(Cotes to William Jones, 20 September 1711 [4], Letter CIV.)

* * *

And again:

You need not give yourself the trouble of examining all the calculations of the Scholium. Such errors as do not depend upon wrong reasoning are of no great consequence & may be corrected by the reader.

(Newton to Cotes, 15 June 1710 [4], Letter XV.)

Throughout the years 1709 to 1713, Cotes continued to exercise his quite exceptional care and thoroughness until the publication of the book in mid-June 1713. It was, of course, printed at the Cambridge

University Press, which Bentley had done so much to restore; 750 copies were printed and Bentley made some personal financial gain (see [5], p. 417). Not so Cotes: not only were his four years of work entirely unpaid; it also went unrewarded by any publicly or, so far as we know, privately, expressed word of thanks or gratitude from Newton. A draft Preface by Newton (from D. T. Whiteside) is cited in [5], p. 114, and contains the statement, translated by Hall and Tilling as follows: 'In publishing all this, the very learned Mr Roger Cotes, Professor of Astronomy at Cambridge, has been my collaborator: he corrected the errors in the former edition and advised me to reconsider many points. Whence it came about that this edition is more correct than the former one.' This small tribute was withdrawn and never printed. It is dated tentatively 'Autumn 1712' and its withdrawal has been widely interpreted by Newtonian scholars to evidence a breach in the relationship between Cotes and Newton. It seems at least plausible that such a breach could stem from the error in the first edition, unremarked by either Newton or Cotes, until reported to Newton by Nicolas Bernoulli on a visit to London in September/October 1712. The error occurs in Proposition x, Book II, on the resistance to the motion of a body in air when the path is a circle. The error had been discovered by Jean Bernoulli (uncle of Nicolas), and the relevant sheets of the second edition, the error still persisting, had been printed off some while before Newton learned of the error. Newton's response was to thank Jean Bernoulli for the information, send him a copy of his *De Analysi* (which William Jones had published (London, 1711)) and successfully propose him as a member of the Royal Society (see [4], p. 142). If Newton's corresponding response to Cotes was to withdraw the compliment in the proposed Preface, and thereafter to fail to make any public acknowledgement of Cotes' great service, then it seems tetchy and ungenerous. Some tetchiness is perhaps understandable, but Newton was not, in general, ungenerous. However, the second edition was nearing completion; both men were anxious to complete the work. During 1712, Cotes was increasingly busy with his own papers and Newton, who was approaching his seventieth birthday, probably found Cotes' persistent attention to detail irritating. Even so, the discovery of a quite important error, so far overlooked by *both*, seems insufficient justification for such a response: but, it must be seen in the light of the generally strained relationships stemming from the Newton–Leibniz controversy and the acrimony which it engendered. It was not that the error itself was so important – it could easily be set right – but that the error was found by Jean Bernoulli who had already reported it to Leibniz and to the French Academy. In his correspondence with Cotes, Newton is matter of fact:

There is an error in the tenth Proposition of the second Book, prob. III, wch will require the reprinting of about a sheet & an half. I was told of it since I wrote to you and am correcting it. I will pay the charge of reprinting it, & send it to you as soon as I can make it ready.

(Newton to Cotes, 14 October 1712 [4], p. 142.)

I mentioned also an error that I was lately told of & wch wants to be set right. I have heard nothing from you this month or above & should be glad of a line to know in what forwardness the Press is.

(Newton to Cotes, 21 October 1712 [4], p. 143.)

You mentioned an Error in the xth Proposition of the II^{d} Book, which will require the reprinting of about a Sheet & an half. I have not revis'd that Proposition to see if I might find it out, but shall stay for Your corrections.

(Cotes to Newton, 23 October 1712 [4], p. 144.)

I send you enclosed the tenth Proposition of the Second book corrected . . . . When this sheet & a quarter is printed off I hope your trouble of correcting will be at an end.

(Newton to Cotes, 6 January 1713 [4], p. 145.)

I have considered Your alteration of Prop. x. Lib. II. and am well satisfied with it . . . Some things in Your Paper I have altered, they are not worth Your notice, being only faults in transcribing.

(Cotes to Newton, 13 January 1713 [4], p. 146.)

Cotes almost invariably signed his letters to Newton, 'Your most Humble Servant', but this last is an exception in being signed 'Your Obliged Freind & Humble Servant' – ironic?; almost certainly not. Some have detected a cooling in the tone of the correspondence after the episode of the undetected error. Personally, I find the tone of the letters from which the above extracts are taken, and of the six further letters which constitute the Cotes–Newton correspondence up to the time of the publication of the second edition, as businesslike and to the point, neither more so nor less, as most of the earlier letters.

Whatever Newton's attitude to Cotes may have been at this stage, six months after the publication of the book, Newton thoughtlessly and indeed discourteously sent to the printer, without reference to Cotes, a list of corrigenda and addenda. Cotes received it via the printer, and wrote to Newton with some asperity. After pointing out that several of the corrigenda were needless or inaccurate, he said of the errata:

I observe You have put down about 20 Errata besides those in my Table. I am glad to find they are not of any moment, such I mean as can give the

reader any trouble. I had myself observ'd several of them, but I confess to
You I was asham'd to put 'em in the Table, lest I should appear to be too
diligent in trifles. Such Errata the Reader expects to meet with, and they
cannot well be avoided. After You have now Your self examined the Book
& found these 20, I beleive You will not be surpriz'd if I tell You I can send
You 20 more as considerable, which I have casually observ'd, & which seem
to have escap'd You: & I am far from thinking these forty are all that may
be found out, notwithstanding that I think the Edition to be very correct. I
am sure it is much more so than the former, which was carefully enough
printed; for besides Your own corrections & those I acquainted You with
whilst the Book was printing, I may venture to say I made some Hundreds,
with which I never acquainted You.

(Cotes to Newton, 22 December 1713 [4], p. 167.)

Thus, this famous correspondence, which began warmly enough and
continued so fruitfully for nearly four years (with a gap from October
1709 to April 1710), ended on a rather chilly note.

Although Cotes' work as editor passed without public recognition,
he gained some reputation from his Editor's Preface. He had gen-
erously (and curiously) offered to acknowledge, as his own, any preface
which Newton and Bentley might agree on between them but, after
discussing it with Newton, Bentley advised Cotes to write it himself,
and not to be so modest. This was a kind of tribute to Cotes, and
possibly Newton regarded it in that light. The correspondence relating
to the Preface is dated March 1712/13, when the dust of the 'undetected error' episode had settled. Cotes then (18 March 1713) outlined
to Newton the form he thought the Preface should take. First, there
would be an account of the book and its improvements. Then a
discussion of the 'manner of philosophising' (that is, a discussion of
the principles of Newtonian philosophy) and how it differed from
that of the Cartesians and others. Cotes would illustrate this with the
deduction of the principle of gravity from the phenomena of nature,
but here there was a difficulty. That a body A attracted a body B
could be deduced from observation by observing B move towards A,
but that body B equally attracted body A looked like a hypothesis (in
Newton's rather restricted sense of 'that which is not deduced from
phenomena') and until this was cleared up, it looked as though Newton
did indeed 'hypothesim fingere'. Next, the Preface would then answer
criticisms which had been raised against the book, such that it deserted
mechanical principles and made use of occult qualities. Finally, there
would be a description of the mathematical principles on which the
book was founded (the method of fluxions). Leibniz would not be
mentioned by name, but Cotes would like permission to quote from
the *Commercium Epistolicum* (The Royal Society (London, 1713)) and

give 'the very words of the Judgement of the Society' (Cotes to Newton, 18 February 1712/13 [4], Letter LXXX). This proposal is not the work of a man who has anything but the highest regard and firmest loyalty to Newton.

In the event, the Preface was scaled down to something less ambitious. In it, Cotes dismissed Peripatetics and Cartesians in a few crisp sentences, and then restricted himself to an account of 'the method of philosophising': 'There is left then the third class, which possess experimental philosophy . . . They frame no hypotheses, nor receive them into philosophy, otherwise than as questions whose truth may be disputed' (R. Cotes in I. Newton, *Principia*, second edition (Cambridge, 1713), Preface). He incorporated Newton's explanation of the difficulty about mutual attraction; argued at length, and cogently, that gravity was deduced from observation of phenomena alone, and answered the objections raised by vorticists and others. In this long Preface, an essay of some seven or eight thousand words, Cotes achieved a clear exposition of Newtonian principles, comprehensible to the eighteenth-century reading public. Modern opinions differ as to the accuracy with which Cotes represented Newton's views: in particular, did Newton hold that gravity was an innate property of matter?; and did the Newtonian theory postulate 'action at a distance'? Cotes' Preface implies that the answer to both these questions is 'yes', whereas most Newtonian scholars would now agree that the answer is 'no'. The Preface was reprinted in English, in Andrew Motte's English translation of the third edition of *Principia*, and is nowadays easily available in Florian Cajori's two-volume paperback reissue of that work (by the University of California Press, 1962). The following reference is taken from Cajori's appendix: 'Maxwell says "And when the Newtonian philosophy gained ground in Europe, it was the opinion of Cotes rather than that of Newton that became most prevalent . . ." (J. C. Maxwell, *Proceedings of the Royal Institution of Great Britain*, vol. 7, 1873–1875, London, pp. 48, 49.)'

2

Logometria

Cotes published one paper only in his lifetime, Logometria. This appeared in the *Philosophical Transactions of the Royal Society*, vol. 29, no. 338, for March 1714 and, quite possibly, did not reach the public until after Cotes' death on 5 June 1716. We know from his correspondence with William Jones that Cotes was working on the paper in 1711: 'I cannot so easily give You an Idea of my other peice concerning Logarithms but I find room enough in this Page to send You one thing out of it as a curiosity which may be understood independently of the rest' (Cotes to Jones, undated, in reply to Jones to Cotes of 1 January 1711/12 [4], Letters CX and CIX). This one piece was the rectification of the logarithmic curve which appears in the Scholium Generale of Logometria, discussed below. On 25 May 1712, Cotes sent the paper to Newton, with a modest and rather confiding letter:

I sent You by Dr Bentley a small Treatise of my own concerning Logarithms, of which the Title is *Elementa Logometriae* together with the Figures belonging to it. I desire the favour of You to deliver 'em to Mr Livebody to be cut in Wood & to give him Your directions if he meet with any difficulty. I fear You are at this time taken up with other buisness, otherwise I would beg of You to peruse the Treatise. You will find I am there proposing a new sort of Constructions in Geometry which appear to me very easy, simple & general. But I am fearfull of relying upon my own Judgement alone, which possibly in this matter may be too much byass'd. What I think to be right, may to others appear whimsical & of no use & I would not willingly give them the satisfaction of laughing at my Dreams. If You think I should venture to publish it, I shall be glad to know what may want to be corrected or altered either in the Matter or Expression. I have been force'd to use some new Terms, as Modulus, Ratio modularis, &c. If others more proper occur to You upon reading the Papers, I shall be very willing to make any alteration. I hope You will pardon this Trouble I give You.

(Cotes to Newton, 25 May 1712 [4], Letter LIV.)

Newton replied promptly enough: 'I received yor papers by Dr Bentley and have run my eye over them. I intend to read them over again & get the cuts done for you as soon as I can find out Mr Livebody' (Newton to Cotes, 27 May 1712 [4], Letter LV). It is a pity we do not have Newton's detailed comments, if he ever made any. Logometria was published in the *Philosophical Transactions of the Royal Society* and, since Newton was President of the Royal Society at the time, we must assume he approved. Certainly, he performed Cotes' errand in arranging for the 'Wooden Cutts'; Cotes thanked him for them on 10 August 1712.

In Logometria, Cotes'

Proposition I: *Defined logarithms as measures of ratios; Established the functional equation $f(x^n) = nf(x)$ which those measures satisfy;*

                  Explained how different systems of logarithms were related through the modulus (a scale factor, in fact the logarithm of e, to the base concerned);

                  Calculated the value of e and of $1/e$, giving the results to 12 decimal places.

Proposition II: *Demonstrated a very neat method of computing Briggs' logarithms, avoiding almost all use of series.*

Proposition III: *Showed how to convert logarithms from one system to another.*

Proposition IV: *Explored at great length the relationship of logarithms to various quadratures of the hyperbola.*

Proposition V: *Described the properties of the logarithmic curve.*

Proposition VI: *Showed the logarithmic properties of the logarithmic spiral.*

In illustrative scholia, Cotes'

Proposition I, Scholium 1: *Obtained the series for $\ln[(z+x)/(z-x)]$ (z constant, x variable).*

Proposition II, Scholium 2: *Extended this to the series for $\ln(1+v)$ and $\ln(1-v)$.*

Proposition III, Scholium 3: *Developed the continued fraction for e and $1/e$, for generating rational approximations to these quantities.*

Proposition IV, Scholium: *Applied the results from the hyperbola to vertical ascent and descent in resisting media.*

Proposition V, Scholium: *Investigated the density of the atmosphere at any given altitude.*

Proposition VI, Scholium: *Showed how to calculate the change in longitude for a given change in latitude, along a loxodrome, bearing 045°.*

The paper ends with a long Scholium Generale, in which Cotes gave his solutions to a number of problems of then contemporary interest, the problems being:

The rectification of the parabola, and the archimedean spiral;
The rectification of the reciprocal spiral, and the logarithmic curve;
The rectification of the cissoid, and to find the volume of the cissoidal solid of revolution;
The area of a conchoid (between the branches), and to find the volume of the conchoidal solid of revolution;
Further quadratures of the hyperbola, and centres of gravity;
The surface area of the hyperboloid of revolution;
The surface areas of the prolate and of the oblate spheroids, and the corresponding volumes;
A note on cubic equations, noting that the solutions depend either on ratios or on angles;
The gravitational attraction of spheroids;
Spiral central orbits described under the inverse cube law;
The oscillation of a cycloidal pendulum in a resisting medium;
The density of air at any altitude, allowing for gravitational variation;
The division of the meridian on the Mercator chart.

This is an impressive list for a first tentative publication. We know that Cotes read widely in the current literature; references to, for example, Briggs, Huygens, Leibniz, Halley, De Moivre, Newton, Wallis, occur in his writings. J. E. Hofmann goes further and says of Logometria: 'Cotes completely worked through and used the whole rich contemporary literature' ([7], p. 368, note 41).

The problems dealt with are not new; the interest lies in the solutions, which Cotes gives, and in the methods of arriving at those solutions, which Cotes does not give. An English translation of Logometria is given in Appendix 1, and it will be seen from this that the solutions are, for the most part, given in geometrical form, although usually requiring the drawing of a line equal in length to a given logarithm or arc length. These lengths have first to be calculated, and Cotes had developed methods of integration applicable to a wide range of problems, enabling the calculations to be performed relatively simply. In Cotes' posthumous works, published as *Harmonia Mensurarum* (Cambridge, 1722), by Robert Smith, Cotes' cousin, successor in the Plumian chair and literary executor, details of the integrals appear as Logometria, Part II, and a further selection of problems, in which the methods are explained in detail, forms Logometria, Part III. These form the subject of the next chapter.

Cotes was engaged on extensions of this work, and making preparations for publishing it, at the time of his death in 1716.

Proposition I of Logometria: *To find the measure of any ratio whatever*, requires, for its full understanding, some knowledge of earlier attempts, in particular that of Halley (to whom Logometria is dedicated), to establish a satisfactory theory of logarithms. Cotes' measure of a ratio is the logarithm of that ratio. Edmund Stone, in general an admirer of Cotes' work, wrote in *A New Mathematical Dictionary* (London, 1743), entry under 'Logarithm': 'Mr Cotes has done this thing in imitation of Dr Halley, although more short, and yet with the same obscurity, for I appeal to anyone, even of his greatest admirers, if they know what he would be at in his first problem ... without having known something of the matter from other principals.' Cotes, of course, was writing for contemporary mathematicians, and not, like Stone, for the more general reader; he could reasonably assume 'some knowledge of the matter' in his readers.

In Proposition I, Cotes argues that, in a ratio $(1+x):1$, x will serve to measure the ratio, and in the sense that this is so, nx will serve to measure the n-fold ratio $(1+x)^n:1$ and, indeed, that any quantity proportional to x will serve as well. A brief examination of the background will help in understanding the argument.

We note in passing, that in any geometric sequence, 1, x, x^2, $x^3, \ldots, x^n, \ldots$, the indices are in arithmetic sequence, and will serve as logarithms of the corresponding terms, and that, in general, whenever the terms of an arithmetic sequence are related one-to-one and in order, to the terms of a geometric sequence, we have a logarithmic relation. In 1695, Halley used this idea as the basis of an analytical definition of logarithms (see [6]). Let a length $AB = 1 + a$, be divided at P_1, P_2, $P_3, \ldots, P_n, \ldots, B$ (Fig. 3), so that AP_1, AP_2, $AP_3, \ldots, AB$ form a geometric sequence, with $AP_1 = 1$ and $P_1P_2 = x$. Then

$$AP_2:AP_1 = (1+x):1; \qquad AP_3:AP_1 = (1+x)^2:1$$

and

$$AB:AP_1 = (1+x)^n:1.$$

Let a length $CD = 1 + b$, be similarly divided at points Q_1, $Q_2, \ldots, Q_m$. Then

$$CD:CQ_1 = (1+x)^m:1.$$

The quantity x, the same in both cases (called by Halley the *unit ratiuncula*), approaches zero, and n, m approach infinity. The limiting ratio $n:m$ is defined as the ratio of the logarithms of AB and CD. Notice that, for large n, the terms of the geometric sequence become

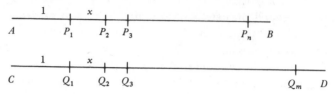

Fig. 3. Halley's *unit ratiuncula*. $P_1P_2 = Q_1Q_2 = x$.

almost equal, therefore their differences become almost equal (the first differences of terms in geometric sequence are also in geometric sequence, and with the same common ratio). These differences P_1P_2, P_2P_3, and so on, form, in the limit, an arithmetical sequence associated with a geometric sequence. We cannot take n as the logarithm of $(1+a)$ because n is infinitely large, but the limiting value of the ratio $n:m$ is finite, and will measure the ratio of the logarithms of AB and CD.

Halley's next step is to let n and m be the same (infinite) number, and then it is the size of x in each case, which will give us the ratio of the logarithms. The length x is clearly $\sqrt[n]{(1+a)} - 1$. Cotes develops Halley's idea, letting PQ be as small as possible, hence arriving at terms PQ_1, PQ_2, and so on, which are approximately in arithmetic sequence. Robert Smith, in his Editor's Notes in *Harmonia Mensurarum*, comments: 'Quo posito haec est vis totius argumentationis.' And so we arrive at the rule of Proposition I, that if $PQ_1 = x$, and measures the ratio AQ/AP, then when the number of points of division gets very large, $PQ_2 = 2x$ and measures the ratio $AQ_2:AP$, $PQ_3 = 3x$ and measures the ratio $AQ_3:AP$, and so on; and in Cotes' words: 'The sum of all the ratios [which means, confusingly, but customary at the time, the product of all the ratios] $(AQ:AP)^n$ is measured by the sum of all the measures, $=nPQ$.' J. E. Hofmann puts it thus: 'If we associate with the ratio $AQ:AP$, the quantity $PQ:AP = (AQ:AP) - 1$, then we must associate with the ratio $(AQ:AP)^n$ the quantity $(AQ:AP)^n - 1 = (1+PQ/AP)^n - 1$, which is approximately equal to $n(PQ/AP)$, if PQ is small' ([7], p. 369), and is 'certain that Cotes intended something of this sort'.

In the corollaries, the idea of a modulus, i.e., a scale factor, is described. Instead of using x in the above analysis, any quantity proportional to x will serve, and different systems of logarithms will result from the choice of different scales, or moduli. If the modulus is 1, the system is that of natural logarithms. Clearly, the logarithms are proportional to the moduli (Corollary 3). In the interesting Corollary 4 to Proposition I and its associated Scholium we have one of the earliest, possibly the earliest, printed value of e. Starting from m as the measure of the ratio $(1+v):1$ to modulus M, i.e.,

$$m = M\left(v - \frac{v^2}{2} + \frac{v^3}{3} - \cdots\right),$$

Cotes solves this for v by reversion of series, to find

$$v = \frac{m}{M} + \frac{m^2}{2M^2} + \frac{m^3}{6M^3} + \cdots,$$

and puts the question: 'What is the value of v so that the measure of $(1+v):1$ shall be equal to the modulus of the system?' This is clearly obtained by putting $m = M$ in the above expression, and we have, by adding 1 to the result,

$$(1+v):1 = \text{the ratio of } 1 + 1 + \tfrac{1}{2} + \tfrac{1}{6} + \tfrac{1}{24} + \cdots \text{ to } 1$$

is a ratio whose measure, in any system whatever, is always equal to the modulus of that system. Thus the modulus emerges as $\log e$ in the given system. The modulus then of natural logarithms is one, and of Briggs', or common logarithms, is $\log e = 0.434 \cdots$. This 'fixed and immutable ratio' is called by Cotes the modular ratio (*ratio modularis*). Natural logarithms of numbers were less readily available in published form than Briggs' logarithms at the time Cotes was writing, and the result $1/0.434 = $ natural log/Briggs' log was frequently used.

In Scholium 3 (see Appendix 1) we find what is almost certainly the first application of a continued fraction to the generation of rational approximation to e. The successive convergents of a continued fraction are, as was clearly known to Cotes, alternately greater than and less than the limit, and Cotes arranges the results in two columns accordingly. Cotes derives the continued fraction by first applying the euclidean algorithm to the ratio $2.718\,281\,828 \cdots : 1$ to generate the quotients $2, 1, 2, 1, 1, 4, 1, 1, 6, 1, 1, 8, \ldots$, and his long explanation, with unaccustomed prolixity, shows how to calculate the successive convergents. For good measure, Cotes adds some intermediate terms, by replacing the multiplications by repeated additions. The arrangement of the work in two columns, one set monotonically increasing, the other monotonically decreasing, is reminiscent of James Gregory's work on the circle, which led him to postulate the existence of another kind of number, resulting from a 'sixth process' – i.e., other than the rational processes and root extraction (see [10], p. 470). J. Wallis had developed his earlier work on continued fractions (*Opera Mathematica*, vol. I (3 vols., Oxford, 1693)), and Brouncker had given a continued fraction for $4/\pi$. Christian Huygens, in *Descriptio Automati Planetarii* (The Hague, 1695), used a continued fraction for π, to obtain rational approximations for use in the calculation of gear ratios. Once again, we see Cotes as widely read, and able to develop the work. The successive approximations are $2:1$, $6:2$, $8:3$, $11:4$, $76:28$ and so on. The convergence is fairly slow. We get three-decimal-place accuracy at $106:39$ and eight-place agreement at $23\,225:8544$.

   In Proposition II: *To construct the Briggsian canon of logarithms*, we find what the reviewer of *Harmonia Mensurarum* in the *Philosophical Transactions of the Royal Society* (probably Robert Smith) called a 'concise uncommon method' (see *Philosophical Transactions of the Royal Society*, 32 (1722), 141). The following explanatory notes will make it clear what Cotes is doing in this proposition (given in Appendix 1): The ratios,

$$\frac{126}{125} = \frac{2 \times 3^2 \times 7}{5^3},$$

$$\frac{225}{224} = \frac{3^2 \times 5^2}{2^5 \times 7},$$

$$\frac{2401}{2400} = \frac{7^4}{2^5 \times 3 \times 5^2},$$

$$\frac{4375}{4374} = \frac{5^4 \times 7}{2 \times 3^7},$$

are chosen to yield rapidly convergent series. Thus

$$\ln \frac{126}{125} = 2\left(\frac{1}{251} + \frac{1}{3(251)^3} + \frac{1}{5(251)^5} + \cdots\right),$$

and similarly for the others. The expansion is in powers of x/z (in the notation of Proposition I, Corollary 5), and in each case x is 1 with z quite large, for example, for the fourth ratio $x = 1$, $z = 8749$. Each number has prime factors, all of which are less than 10. To calculate $\ln 10$, it is then necessary to combine these four ratios so that the result is $\ln 2 + \ln 5$. Let

$$\ln 10 \text{ be } a \ln \frac{126}{125} + b \ln \frac{225}{224} + c \ln \frac{2401}{2400} + d \ln \frac{4375}{4374},$$

giving

$$\ln 10 = a(\ln 2 + 2 \ln 3 + \ln 7 - 3 \ln 5) + b(2 \ln 3 + 2 \ln 5$$

$$- 5 \ln 2 - \ln 7) + c(4 \ln 7 - \ln 3 - 2 \ln 5 - 5 \ln 2)$$

$$+ d(4 \ln 5 + \ln 7 - \ln 2 - 7 \ln 3)$$

$$= (a - 5b - 5c - d) \ln 2 + (2a + 2b - c - 7d) \ln 3 + (-3a + 2b$$

$$- 2c + 4d) \ln 5 + (a - b + 4c + d) \ln 7.$$

Putting now the coefficients:

of ln 2 = 1,

of ln 3 = 0,

of ln 5 = 1,

of ln 7 = 0,

we have the following set of linear equations:

$a - 5b - 5c - d = 1$,

$2a + 2b - c - 7d = 0$,

$-3a + 2b - 2c + 4d = 1$,

$a - b + 4c + d = 0$,

the solutions of which are Cotes' coefficients:

$a = 239$, $b = 90$, $c = -63$, $d = 103$.

This lengthy procedure avoids having to use the twelve or so terms of the series given in Proposition I, Corollary 5, to achieve twelve-decimal-place accuracy.

Multiplication by the modulus of Briggs' logarithms will convert the logarithms so calculated (modulus 1, i.e., base e) to Briggs' logarithms. Cotes appeals to the proportionality between logarithms and the moduli of their systems (or canons) set out in Proposition I, Corollary 3, to calculate the modulus of the Briggs canon. Other values of a, b, c, d are selected to yield, in turn ln 7, ln 5, and ln 3 and multiplication by the Briggs modulus gives the Briggs logarithm. To calculate logarithms of primes greater than 10, for example log P. Cotes' verbal explanation can be symbolised as follows:

$$(P+1)(P-1) = P^2 - 1, \quad \text{so} \frac{P^2}{(P+1)(P-1)} \quad \text{is of the form} \frac{N+1}{N}.$$

Proposition I, Corollary 5, can thus be applied to give a rapidly convergent series and

$$\ln \frac{P^2}{(P+1)(P-1)} + \ln (P+1)(P-1) = \ln P^2,$$

therefore, ln P is obtained on division by 2. Note that $(P+1)(P-1)$ has prime factors less than P, hence its logarithm is known. Thus, complete tables can be constructed, such as those of Briggs and Vlacq. For interpolation or extrapolation of these tables Cotes suggests that the first term only of the series (Corollary 5) is usually sufficient, and the expressions given in the Corollary to Proposition II result from doing just this.

Proposition III is: *To show the construction of a canon of logarithms by any system whatever.* Given a system of ratios and their measures, a system of logarithms can be adapted, or constructed (or calculated) to suit it. Such a system of ratios and their measures is called by Cotes 'logometric'. The remaining three propositions in Logometria demonstrate three curves which yield logometric systems, and apply the ideas developed so far to the solution of three physical problems. The curves selected are the hyperbola, the logistica (or logarithmic curve) and the equiangular spiral. The logarithmic properties of these curves had been fully investigated during the preceding century. The problems to which they are applied were equally well known, namely, the motion under gravity of a heavy body in a resisting medium, the density of the atmosphere at any given height, and the change in longitude corresponding with a given change in latitude along a rhumb line (line of constant bearing). Cotes is concerned to show that his logarithmic methods lead to simple solutions.

We now come to the all-important Proposition IV: *The quadrature of any hyperbolic space by a canon of logarithms*, in which the logarithmic properties of hyperbolic spaces are investigated. The results are related to Cotes' work on integrals, and the whole armoury deployed against a wide range of problems in the Scholium to Proposition IV and in the Scholium Generale. In any detailed study of Cotes' work, Proposition IV merits attention, and so a detailed commentary is given in Appendix 2.

The example chosen to illustrate the results is the motion of a heavy body ascending and descending, vertically, in a medium whose resistance varies as the square of the velocity. The problem is discussed in detail in *Principia*, second edition (Cambridge, 1713), Book II, Propositions VIII and IX, as acknowledged by Cotes. The related Problem X (circular motion in a resisting medium) contained the famous error pointed out to Newton by Nicolas Bernoulli. The interest to Cotes of Proposition IV is that the different aspects of the problem lead to differential equations whose solutions are either trigonometric or logarithmic. The solutions are expressed in geometric form, and beautifully summarised in terms of the properties of the circle and of the hyperbola.

In Propositions V: *To describe the logarithmic curve by a canon of logarithms*; and VI: *To adapt a canon of logarithms to the equiangular spiral*, and their Scholia, Cotes considers two problems of great con-

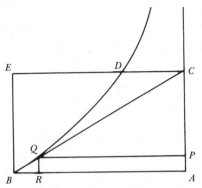

Fig. 4. BQD is the logarithmic curve, with constant sub-tangent AC. From *Harmonia Mensurarum*, Cambridge University Press, 1722.

temporary interest, namely, atmospheric pressure and its relation to gravitational attraction, and the navigational problem of determining longitude at sea. Once again, he has obviously read widely in the fields and proposes some elegant solutions in logarithmic form.

In Proposition v, Cotes states the defining property of the logarithmic curve, namely, $PQ = \ln (p:q)$ and from this obtains methods of constructing the curve (and hence of finding logarithms) from given data. The constancy of the sub-tangent follows fairly simply, thus (Fig. 4), $AP \propto \ln (AB:PQ) = \ln (AB:AR) \propto (BR:AR)$ or $(BR:AB)$, by Proposition i, $= (AP:AC)$ by similar triangles, i.e., $AP \propto (AP:AC)$, i.e., AC is constant. Clearly, the value of $\ln (p:q)$ will depend on the scale adopted, and this will be reflected in the (constant) length of the sub-tangent. The length of the sub-tangent thus fits nicely into Cotes' scheme, as the modulus. This kind of geometrical fluxion technique is also used to show that the infinitely extended area between the curve and its asymptote, measured from a given point, has a finite value. This was not an entirely new idea at the time. Cotes wants to use it in the Scholium to Proposition v to show that the weight of an infinite atmosphere is finite.

The application of the properties of the logarithmic curve to the investigation of the pressure of the atmosphere is set out in this scholium. The problem is discussed in general terms in *Principia* (using properties of the hyperbola) in, for example, the third edition (Cambridge, 1726), Book ii, Proposition xxii, Problem xvii; and in the second edition, p. 470, we find 'the density of the air at a height equal to the radius of the earth' computed. Cotes would have known of Gregory's work in this area (see [10], p. 420) and of the practical investigations carried out in the seventeenth century. The following extract from one of Cotes' *Hydrostatical and Pneumatical Lectures*, ed. R. Smith, second edition (Cambridge, 1747), p. 111, delivered to undergraduates, is interesting, not only for its content, but also for its style – no mention of logarithmic curve here!

We might have made with the Torricellian tube an experiment like this, to shew the different pressure of the air at different distances from the surface of the earth, had the Observatory been much higher than it is. At the altitude of 54 feet the ascent of the quicksilver would be too small to ground anything upon, being about 1/20 of an inch: It was therefore necessary to make use of the contrivance you have seen,* to supply the defect of some very high mountain, upon which had any such been near us, we might have observed a sensible alteration, even with the Barometer. Such an experiment was formerly made at the desire and by the direction of Mr Pascal, in the year 1648, upon the Puy de Domme, a very high mountain in France. It was then observed that in ascending 3000 Paris feet, the quicksilver in the tube fell down 3 inches and 1/8 of an inch. To reduce this to English measure, we may say that ascending 3204 English feet, the height of the quicksilver was abated 3 inches and 1/3 of an inch. Another experiment like this was made by Mr Caswell upon Snowdon Hill in Wales; he found that the height of 3720 feet abated the quicksilver 3 inches and 8/10. It may not be amiss to add here the result of a computation which I made of the weight of all the air which presses upon the whole surface of the earth. If this weight were to be expressed by the number of pounds it contains, that number would be so large as to be in a manner incomprehensible. I will therefore make use of another way of expressing it, by determining the diameter of a sphere of lead, of the same weight with all the air which presses upon the whole surface of the earth. Now that diameter was found to be very nearly 60 miles long. If any one has a desire to make this calculation after me, he may proceed upon these grounds. That the weight of a column of air reaching to the top of the atmosphere, is most commonly equal to a column of water having the same basis, and the altitude of 34 feet; that the semidiameter of the earth is equal to 20 949 655 feet, and that the specific gravity of water is to that of lead as 1000 to 11 325.

* Cotes seems to have repeated Pascal's experiment to demonstrate the pressure of the air, using glass tubes some forty feet long and water. He says in an earlier lecture: 'I procured the apparatus for that chargeable and troublesome experiment of Mr Pascal, rather as a curiosity than as absolutely necessary to our purpose' (*Ibid*, p. 98).

The curve considered in the final proposition, Proposition vi, of *Logometria*, is again a curve in which elements in arithmetical progression, in this case the polar angles, are linked with elements in geometric progression, the corresponding polar radii. The curve is the logarithmic or equiangular spiral and again this 'exhibits a complete and perfect system of logarithms'. The application to navigational problems arises because the spiral is the stereographic projection

of a line of constant bearing on the sphere – a loxodrome or rhumb
line. The Scholium to Proposition VI is happily chosen, probably as
a tribute to Halley, the then secretary of the Royal Society. Halley
had written on the application of the logarithmic spiral – 'A most
compendius method of constructing the logarithms, exemplified and
demonstrated from the nature of numbers, without any regard to the
hyperbola', *Philosophical Transactions of the Royal Society*, vol. 19 (1695),
no. 215, pp. 58–67 – to 'the problem of the nautical meridian'. Cotes,
as Plumian Professor, was *ex-officio* a member of the Board of Longitude, the committee appointed by Parliament to consider the merits
of proposals for determining longitude at sea. He had therefore more
than a passing interest in this particular problem. Halley, in his paper,
describes the nautical meridian thus: 'The nautical meridian line is a
table of longitudes answering to each minute of latitude on the rhumb
line making an angle of 45° with the meridian. Wherefore the
[nautical] meridian is none other than a scale of logarithmic tangents
of the half complements of the latitudes.' A rhumb line is a line of
constant bearing. The problem was to determine what change of
longitude corresponded with a given change in latitude, for a course
along a given rhumb line. Given reasonable sailing conditions, a
constant course could be held over reasonably short distances, and
the latitude change found from fairly well tried methods, for example
observations of the noon altitude of the sun. Cotes, following Halley,
selects a rhumb line at 45° to illustrate the principles. Thus Cotes
begins by demonstrating (Appendix 1, Fig. 10) that the arc *ab* is the
longitude change corresponding with the latitude change *ag*. By
stereographic projection of the surface of the sphere onto the
equatorial plane, the meridians become straight lines through the
pole, and the rhumb line, from the angle-preserving properties of
the projection, becomes an equiangular spiral. (Halley says that he
had the result on angle preserving from De Moivre, but that Robert
Hooke had claimed he had it sooner.)

Note on stereographic projection
By projection from S on to XY (Fig. 5), P is mapped to P' and A to
A'. If the arc PA subtends an angle θ at the centre of the sphere,

$$P'A' = R \tan \frac{\theta}{2}$$

$$= R \tan \tfrac{1}{2}(90° - \phi),$$

where ϕ is the latitude of A. The auxiliary circle is of radius R, the

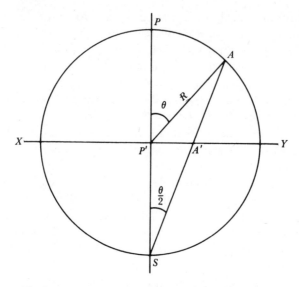

Fig. 5. In stereographic projection, the surface of the sphere is mapped by projection from the pole S onto the equatorial plane through XY.

radius of the earth; and for a 45° rhumb, the point F (Appendix 1, Fig. 9) coincides with the point C. Hence,

$$\text{arc } ab = \log \frac{\tan \frac{1}{2}Pd}{\tan \frac{1}{2}Pa}, \quad \text{to modulus } R$$

$$= R(\ln \tan \tfrac{1}{2}Pd - \ln \tan \tfrac{1}{2}Pa).$$

And, generally, when changing latitude from ϕ_1 to ϕ_2, the corresponding longitude change is

$$\ln \tan \left(\frac{90 - \phi_2}{2}\right) - \ln \tan \left(\frac{90 - \phi_1}{2}\right).$$

For a rhumb line at an angle other than 45°, the point F (Appendix 1, Fig. 10) will not coincide with C, and the modulus of the measures will not be R. That is, the longitudinal changes corresponding with latitude changes, along different rhumb lines, will give rise to logarithms of different moduli. Conversely, logarithms to different moduli will apply directly to different rhumb lines. According to Halley, Vlacq's logarithms give the true meridional parts, at one minute intervals, along a rhumb bearing 51° 38′ 9″.

Finally, Cotes concludes the Scholium to Proposition VI by showing that meridional part $ab = R[\ln \tan (Pa/2) - \ln \tan (Pg/2)]$ and, putting

R in first minutes of arc and using Briggs' logarithms,

$$\text{arc } ab \text{ (in minutes)} = 3437.746\,77\cdots$$

$$\times\frac{1}{0.434}\cdots\left(\log\tan\frac{Pa}{2}-\log\tan\frac{Pg}{2}\right)$$

$$=7915.7044\cdots\left(\log\tan\frac{Pa}{2}-\log\tan\frac{Pg}{2}\right).$$

(Note: The value $3437.746\,77\cdots$ assumes a spherical earth, and the value $3.141\,59\cdots$ for π.)

3

New methods in the calculus

The elegant and economical presentation of the solutions of the problems in the Scholium Generale of Logometria (Part II) conceals the no less economical and elegant methods by which Cotes arrived at these solutions. The methods depended on imaginative use of the tables of integrals which Cotes had constructed, enabling him to handle logarithmic, trigonometric and hyperbolic forms with ease and confidence. We have seen that, quite early in his career, in fact whilst still at school, Cotes had an interest in 'the squaring of curves', and his first letter to Newton (18 August 1709 [4], Letter II) concerned corrections he had made to Newton's fluents as published in *De Quadratura Curvarum* (London, 1704). From Robert Smith's *Editoris Notae* to *Harmonia Mensurarum* (Cambridge, 1722), it is clear that Cotes had developed these tables before 1714. Had they appeared along with the 1714 publication of Logometria, they would have enjoyed rather more than their short vogue, before being inevitably superseded by the superior notation and techniques soon to be developed by the Continental mathematicians.

Working without an established and accepted notation, and before the clear recognition of hyperbolic and inverse trigonometrical functions, Cotes developed eighteen tables of integrals, and associated reduction formulae (called by him 'Continuation Formulae'). By their aid, he was able to attack, and solve successfully, a number of problems involving logarithmic, trigonometric and hyperbolic functions. Previous solutions had depended on 'integration by conic areas', i.e., relating the integrand to the differential of a known conic area, or on term-by-term integration of a (possibly convergent) series. Cotes' methods marked a great improvement. They did not appear in print until the 1722 publication of *Harmonia Mensurarum*, and one agrees readily with De Morgan's remark that they: 'represent the first substantial advance in the development of integration techniques applied to logarithmic and trigonometrical expressions' (*Penny Cyclopaedia*, vol. 8 (1857), p. 87). Some of the problems in the Scholium Generale arise from Cotes' work on *Principia*, second edition (Cambridge, 1713), notably those on spirals, and gravitational attraction of spheroids; others, familiar problems of the day, are well chosen to demonstrate the superiority of the methods, as, for example, the rectification of the parabola and of the logistic (logarithmic) curve. The final problem of the Scholium Generale shows a neat construction for the length of the meridian on the Mercator chart, effectively the integral of sec θ between suitable limits.

It is clear then, that the link between the principles set out in Logometria, and the problems to which they are applied, is provided by the tables of integrals. These are much more than a collection of more or less ingenious integration techniques. They epitomise Cotes' perception of that 'Harmony of Measures' between measures of angles (trigonometrical quantities) and measures of ratios (logarithms): 'That Harmony of Measures, which is so strong that I propose a single notation serve to designate measures, whether of ratios or of angles' (R. Cotes, Logometria, Part II, Preface). It is this 'single notation' invented by Cotes, and the rigid adherence to the Newtonian fluxional notation which gives the tables their rather complicated and, at first sight, forbidding appearance. The tables give the fluents (integrals) of given fluxional forms (loosely speaking, of given differentials). z is the variable; $\dot{z}$ its fluxion; d, e, f, g, η, θ are constants; η is the power of z, and the integrals are tabulated for a limited range of values of θ. This is the same as the notation used by Newton and emphasises Cotes' work as arising from Newton's. Newton had, at one time, advocated the construction of tables of fluents as being a useful service to mathematicians. Cotes, however, expresses the integrals in terms of R, S and T, where these three quantities are the sides of a right-angled triangle, and the symbol $R\left|\dfrac{R+T}{S}\right.$ has two meanings, 'according as R is the square root of a positive or of a negative quantity':

(i) If R^2 is positive, the symbol means 'the measure of the ratio of $R+T$ to S, to the modulus R', i.e., $R \ln \dfrac{R+T}{S}$.

(ii) If R^2 is negative, the symbol means 'the measure of the angle whose radius, sine and tangent are as $R:S:T$', i.e., $R \times \text{arc} \tan T$ (the angle being measured in radians), R being assumed positive, i.e., if, for example, $R = -16$ then R is taken as 4.

Tables I and II are illustrated in Fig. 6. In Table II for $\theta = 0$, for

θ	FORMA I. $\dfrac{d\dot{z}z^{\theta\eta-1}}{e+fz^{\eta}}$.
&c.	
5.	$\frac{dz^{4\eta}}{4\eta f} - \frac{dez^{3\eta}}{3\eta ff} + \frac{deez^{2\eta}}{2\eta f^{3}} - \frac{de^{3}z^{\eta}}{\eta f^{4}} + \frac{de^{4}}{\eta f^{5}} \Big\vert \frac{e+fz^{\eta}}{e}$.
4.	$\frac{dz^{3\eta}}{3\eta f} - \frac{dez^{2\eta}}{2\eta ff} + \frac{deez^{\eta}}{\eta f^{3}} - \frac{de^{3}}{\eta f^{4}} \Big\vert \frac{e+fz^{\eta}}{e}$.
3.	$\frac{dz^{2\eta}}{2\eta f} - \frac{dez^{\eta}}{\eta ff} + \frac{dee}{\eta f^{3}} \Big\vert \frac{e+fz^{\eta}}{e}$.
2.	$\frac{dz^{\eta}}{\eta f} - \frac{de}{\eta ff} \Big\vert \frac{e+fz^{\eta}}{e}$.
1.	$\frac{d}{\eta f} \Big\vert \frac{e+fz^{\eta}}{e}$.
0.	$\frac{-d}{\eta e} \Big\vert \frac{e+fz^{\eta}}{fz^{\eta}}$.
−1.	$\frac{-d}{\eta ez^{\eta}} + \frac{df}{\eta ee} \Big\vert \frac{e+fz^{\eta}}{fz^{\eta}}$.
−2.	$\frac{-d}{2\eta ez^{2\eta}} + \frac{df}{\eta eez^{\eta}} - \frac{dff}{\eta e^{3}} \Big\vert \frac{e+fz^{\eta}}{fz^{\eta}}$.
−3.	$\frac{-d}{3\eta ez^{3\eta}} + \frac{df}{2\eta eez^{2\eta}} - \frac{dff}{\eta e^{3}z^{\eta}} + \frac{df^{3}}{\eta e^{4}} \Big\vert \frac{e+fz^{\eta}}{fz^{\eta}}$.
−4.	$\frac{-d}{4\eta ez^{4\eta}} + \frac{df}{3\eta eez^{3\eta}} - \frac{dff}{2\eta e^{3}z^{2\eta}} + \frac{df^{3}}{\eta e^{4}z^{\eta}} - \frac{df^{4}}{\eta e^{5}} \Big\vert \frac{e+fz^{\eta}}{fz^{\eta}}$.
&c.	

θ	FORMA II. $\dfrac{d\dot{z}z^{\theta\eta+\frac{1}{2}\eta-1}}{e+fz^{\eta}}$
&c.	
3.	$\frac{2dz^{\frac{5}{2}\eta}}{5\eta f} - \frac{2dez^{\frac{3}{2}\eta}}{3\eta ff} + \frac{2deez^{\frac{1}{2}\eta}}{\eta f^{3}} - \frac{2ee}{\eta f^{3}}dR \Big\vert \frac{R+T}{S}$.
2.	$\frac{2dz^{\frac{3}{2}\eta}}{3\eta f} - \frac{2dez^{\frac{1}{2}\eta}}{\eta ff} + \frac{2e}{\eta ff}dR \Big\vert \frac{R+T}{S}$.
1.	$\frac{2dz^{\frac{1}{2}\eta}}{\eta f} - \frac{2}{\eta f}dR \Big\vert \frac{R+T}{S}$.
0.	$\frac{2}{\eta e}dR \Big\vert \frac{R+T}{S}$.
−1.	$\frac{-2d}{\eta ez^{\frac{1}{2}\eta}} + \frac{2}{\eta e}dr \Big\vert \frac{r+t}{s}$.
−2.	$\frac{-2d}{3\eta ez^{\frac{3}{2}\eta}} + \frac{2df}{\eta eez^{\frac{1}{2}\eta}} - \frac{2f}{\eta ee}dr \Big\vert \frac{r+t}{s}$.
−3.	$\frac{-2d}{5\eta ez^{\frac{5}{2}\eta}} + \frac{2df}{3\eta eez^{\frac{3}{2}\eta}} - \frac{2dff}{\eta e^{3}z^{\frac{1}{2}\eta}} + \frac{2ff}{\eta e^{3}}dr \Big\vert \frac{r+t}{s}$.
&c.	

$$R = \sqrt{\frac{-e}{f}}, \quad T = z^{\frac{1}{2}\eta}, \quad S = \sqrt{\frac{e+fz^{\eta}}{f}}.$$

$$r = \sqrt{\frac{-f}{e}}, \quad t = \frac{1}{z^{\frac{1}{2}\eta}}, \quad s = \sqrt{\frac{e+fz^{\eta}}{ez^{\eta}}}.$$

Fig. 6. Cotes' tables of fluents. From *Harmonia Mensurarum*, Cambridge University Press, 1722.

example, we are concerned with evaluating the fluent of

$$\frac{d\dot{z}\, z^{\frac{1}{2}n-1}}{e+fz^{n}},$$

i.e., replacing Cotes' d with 1 to avoid confusion, the table gives

$$\int \frac{1z^{\frac{1}{2}n-1}}{e+fz^{n}}\, dz,$$

and this has two forms according as e and f have like or unlike signs. The integrals are, after substitution and a little reduction;

$$\frac{2}{\eta e}\sqrt{\frac{e}{f}}\arctan\frac{z^{\frac{1}{2}n}}{\sqrt{\frac{e}{f}}}$$

or

$$\frac{2}{\eta e}\sqrt{\frac{e}{f}}\ln\frac{\sqrt{e/f}+z^{\frac{1}{2}n}}{\sqrt{e/f}-z^{n}}$$

and, on substituting the values of R, S and T as given in Fig. 6, these agree with Cotes' results, as follows. Consider the two integrals,

$$\int \frac{z^{\frac{1}{2}n-1}\, dz}{e-fz^{n}}=\frac{2}{\eta e}\sqrt{\frac{e}{f}}\ln\frac{\sqrt{e/f}+z^{\frac{1}{2}n}}{\sqrt{\frac{e-fz^{n}}{f}}}, \tag{1}$$

$$\int \frac{z^{\frac{1}{2}-1}\, dz}{e+fz^{n}}=\frac{2}{\eta e}\sqrt{\frac{e}{f}}\arctan\frac{z^{\frac{1}{2}n}}{\sqrt{e/f}}. \tag{2}$$

Cotes gives the single result,

$$\text{fluent of }\frac{z^{\frac{1}{2}n-1}\dot{z}}{e+fz^{n}}=\frac{2}{\eta e}\,R\left|\frac{R+T}{S}\right.,$$

where

$$R=\sqrt{\frac{-e}{f}}, \qquad T=z^{\frac{1}{2}n}, \qquad S=\sqrt{\frac{e+fz^{n}}{f}},$$

the two cases being distinguished by the two interpretations to be put on the notation. Replace f with $-f$ in (1) and (f now being a negative number) we see that this agrees with Cotes' result, except for the additive constant

$$\frac{2}{\eta e}\sqrt{\frac{-e}{f}}\ln\frac{1}{\sqrt{-1}}.$$

As Smith says in his Editor's Notes, in a different context: 'Whatever the measure of $\sqrt{-1}$ may be, it is at least a constant.' In (2) the result agrees with Cotes' form, so long as we remember that the complex nature of R is only the pointer, to indicate that the notation is to be read as an angle. In computing the angle, the complex nature of R is ignored. This dual interpretation of the single notation implies that if we replace R with iR in the logarithmic interpretation, we shall arrive at the trigonometrical interpretation, i.e., that $iR \ln \dfrac{iR+T}{S}$ differs by, at most, a constant from $R \times \theta$, where θ is the angle whose radius, secant and tangent are as R, S and T. Putting then

$$iR \ln (\sin \theta + i \cos \theta) = R\theta + C,$$

i.e.,

$$\ln (\sin \theta + i \cos \theta) = -i\theta + D,$$

the correct result follows by choosing $D = \tfrac{1}{2}i\pi$, which gives

$$\ln (\cos \phi + i \sin \phi) = i\phi, \text{ where } \phi = \tfrac{1}{2}\pi - \theta.$$

Thus, Cotes' dual notation is a statement of this result, although it never seems to have been clearly formulated by Cotes, except briefly and in passing in Logometria (*Harmonia Mensurarum*, p. 28), where the statement is equivalent to

$$i \ln (\cos \theta + i \sin \theta) = \theta,$$

which has a sign error. (This error was pointed out by I. Schneider in an article 'Der Mathematiker Abraham De Moivre', *Archive for the History of Exact Science*, vol. 5 (31 December 1968), nos. 3 and 4, p. 236.) This result lies at the heart of Cotes' thinking, it is his true *Harmonia Mensurarum*, but nowhere in his surviving papers does he discuss it in detail. However, two of his commentators, namely, Robert Smith (as editor) and Nicholas Saunderson (of whom more later), neither of them aware of the significance of the result, offer proofs which amount to demonstrating that

$$iR \ln \frac{iR+T}{S} \quad \text{and} \quad R \times \arc \tan \frac{T}{R},$$

have the same fluxion, which is $(-R^2/S^2)\dot{T}$. This is easily proved, subject to the condition that $(iR)^2 + T^2 = S^2$ in the first case, and $R^2 + T^2 = S^2$ in the second case. Both commentators refer to the principles developed in Logometria for the first, and offer a neat little piece of 'geometrical analysis' for the second.

In his Preface to Logometria, Part II, Cotes introduces the concept of the measure of an angle. There, he considered that enough had been said about the measures of ratios but a word was needed about angles. The circular arc, intercepted between the legs of an angle and having its centre at the point of the angle, was the natural measure; but this would vary according to the size of the circle, therefore a modulus was needed. The modulus could be a standard circle, or some standard line which varied as the circle, for example the diameter, the side of a regular polygon, and so on. The radius was the most suitable choice because, whenever the measure of a ratio was changed into the measure of an angle, the modulus could be changed into the radius. Robert Smith expands the idea in his Editor's Notes. Following closely the form of the argument in Cotes' Proposition I in Logometria, Part I, where the modular ratio is shown to be 2.718 28 $\cdots$ (i.e., the ratio whose measure is always equal to the modulus), Smith arrives at the idea of a modular angle, i.e., an angle whose measure is always equal to the radius, and shows it to be 57.295 degrees. This is probably the first published calculation of one radian in degrees. Smith first derives the series for arc sin m, where m is the measure of an angle, uses Newton's method of reversion of series to find a series for m, and then puts m equal to the chosen modulus M. In this way he finds the series for the sine of the angle whose measure is equal to the modulus – the modular angle. It is analogous, step by step, to Cotes' determination of the modular ratio in Logometria, Proposition I. Smith says he found it in a small paper of Cotes'; unfortunately, this paper has not survived.

The complete set of eighteen forms of fluents worked out by Cotes is shown in Appendix 3, together with the values of θ for which they were tabulated, and the associated reduction formulae for extending the tables. The value of each fluent for $\theta = 0$ is given in Appendix 4. Readers might enjoy interpreting some of the fluents in modern form, as in the following:

Form IX for $\theta = 1$ will serve as an illustration. Fluxional Form IX is

$$\frac{d\dot{z}\,z^{n-1}}{(g+hz^n)\sqrt{(e+fz^n)}} \quad \text{(if } \theta \text{ is put} = 1)$$

and the fluent is given as

$$\frac{2}{\eta(fg-eh)}\,d\mathrm{R}\left|\frac{\mathrm{R}+\mathrm{T}}{\mathrm{S}}\right.,$$

where

$$\mathrm{R} = \sqrt{\frac{eh-fg}{h}}, \qquad \mathrm{T} = \sqrt{e+fz^n}, \qquad \mathrm{S} = \sqrt{\frac{fg+hz^n}{h}}.$$

To obtain this result, substitute $e + fz^n = w^2$, as Cotes probably did, and we obtain

$$\frac{2d}{\eta} \frac{\dot{w}}{(fg - eh) + hw^2},$$

a fluxion of Form II (see Fig. 6). This has two forms according as $fg - eh$ is positive or negative. If positive, we would today integrate the form to give

$$\frac{2d}{\eta\sqrt{h}} \frac{1}{\sqrt{fg - eh}} \arctan \sqrt{\frac{(e + fz^n)h}{fg - eh}}.$$

The angle is that defined by Cotes' R, T and S, as given above, so long as we follow the rules:

(i)   If R is the square root of a negative quantity, the fluent is the measure of an angle;

(ii) in computing the angle, the complex nature of R is ignored, it is only a guide as to whether the fluent is the measure of an angle or of a ratio.

The logarithmic form can be checked in the same way. It will be found to agree with Cotes' stated result.

Some more general currency was given to Cotes' integral tables by the publication, in 1730, of Edmund Stone's English translation (*The Method of Fluxions Both Direct and Inverse* (London, 1730)) of the Marquise de L'Hôpital's *Analyse des Infinement Petits* (Paris, 1696). L'Hôpital dealt only with the differential calculus (which he had learned from Jean Bernoulli) saying he understood that Leibniz was about to publish a book on the integral calculus. This book did not appear, and Stone added his own appendix on the integral calculus, firmly adopting a fluxional approach, and advocating the use of Cotes' tables in the following words:

Before I conclude this Section, I will take the Liberty of adding a Word or two concerning the excellent compendious Method of expressing the Fluents of given Fluxions by Measures of Ratios and Angles by the late Mr Cotes, Professor of Astronomy and Experimental Philosophy in the University of Cambridge; and published after his Death by his Successor Dr Smith, under the Title of *Harmonia Mensurarum*.

Here the Labour of throwing Quantities into infinite Series, which in many Cases is very troublesome, and on account of their converging too slowly are not fit for Use, is entirely avoided, and elegant Constructions of the Fluents of Fluxions are had geometrically, with the Assistance of ample Tables of Logarithms of Brigg's Form, for finding the Measures of Ratios, and of large Tables of natural Sines and Tangents for finding the Measures of Angles.

And from here may be deduced wonderful neat and compendious Solutions of all difficult Problems, such as the Quadrature of Curve-lin'd Spaces, Rectification of Curves, Cubation of Solids, &c. wherein the Fluents of given Fluxions are concerned. Severall examples of which I shall give hereafter.

Stone gives two simple numerical examples to explain the notation:

(i) 'To calculate $R\left|\dfrac{R+T}{S}\right.$ where R is the square root of an affirmative quantity.'

$$R = 8, T = 6, S = 10.$$

This is $8 \times \ln 1.4 = 2.691\,67 \cdots$ but Stone uses Brigg's logarithms (more readily available in published form). He uses Cotes' result: 'As the moduli are to each other, so are the measures.' The modulus of Briggs' logarithms is 0.434, the modulus R in the given example is to be 8, hence $0.434 : 8 ::$ measure of 1.4 to Briggs' modulus : measure of 1.4 (this last term means ln 1.4); i.e., $0.434 : 8 :: \log_{10} 1.4 : \ln 1.4$. Hence the required measure is $2.691\,67 \cdots$ as above.

(ii) 'To find the value of $R\left|\dfrac{R+T}{S}\right.$, where R is the square root of a negative quantity, and so impossible.' The example given is

$$R = 16, \qquad T = 12, \qquad S = 20 \ (\text{but } R^2 \text{ is } -256).$$

This is to be interpreted as

$$16 \times \arctan 0.75$$

$$= 16 \times 0.6433 \ (\text{radians})$$

$$= 10.29 \text{ approximately.}$$

Stone's method is to find arc tan 0.75 as $36°\,52'\,6'' = 36.868 \cdots °$, and to use

$$\frac{\text{modulus of trigonometrical canon}}{\text{trigonometrical measure}} = \frac{\text{modulus of system}}{\text{required measure}},$$

i.e.,

$$\frac{57.295 \cdots}{36.868 \cdots} = \frac{16}{\text{required measure}},$$

giving the required measure, i.e., the value of $R\left|\dfrac{R+T}{S}\right.$ as $10.433 \cdots$. (There is a small computational error here in Stone's work.)

These two examples serve to show the nature of the computations required, and the way in which they could be reduced to relatively straightforward working rules, using published tables then available. Stone also gives a simple geometrical construction, which avoids the use of tables, for constructing the measure of the angle in the second case.

A more detailed and useful commentary on Cotes' integrals was given by Nicholas Saunderson (1622/3–1739), the blind successor of Cotes' friend, Whiston, to the Lucasian chair. Saunderson was much revered as a teacher, and as an expounder of Newtonian philosophy; and known, among other things, as the inventor of Saunderson's 'Palpable Arithmetic', a device which enabled him to perform elaborate computations, using his sense of touch. In the Preface to Saunderson's posthumous *The Elements of Algebra in Ten Books* (Cambridge, 1740), his son John wrote:

At this time [1711] the ingenious Mr Cotes filled the Plumian chair of Astronomy and Experimental Philosophy, a man of great sweetness of temper, and engaged to our author in the strictest friendship; of the same age, of the same genius and inclination to the mathematics, both approved and recommended to professorships by Sir Isaac Newton. No University could ever at one time boast of two so capable and so disposed to promote the study of philosophy among her pupils. Had they lived to more mature ages, mutually assisting and inspiring each other in the pursuit of knowledge, what glory might have accrued to our university, what advancement to science from their united labours! But Mr Cotes was hurried away by a fever in the flower of his age, having only time to compose a few pieces as specimens of his extraordinary capacity, but of great value to the learned. And our author's life, though longer, was so devoted to lectures, that he now leaves to posterity as few monuments of his abilities.

In his *Method of Fluxions* (London, 1756), Saunderson derived all Cotes' integrals, using entirely analytical procedures, in a chapter called 'The Cotesian forms of fluents computed'. He derived the integrals one from the other in the order indicated in Appendix 3, Fig. 1. He used simple substitutions, the Cotes reduction formulae, and a device which I have called Saunderson's rule. For example, the substitution $x = e + fz^n$ reduces Form I to $(d/\eta f)(\dot{x}/x)$, whose fluent $\dfrac{d}{\eta f} \left| \dfrac{e + fz^n}{e} \right.$ follows at once from Logometria, Proposition I. The reduction formula applicable to this, and to Form II if η is a fraction whose denominator is 2, is

$$\frac{1}{\theta \eta} dZ^{\theta n} = I_\theta + I_{\theta + 1},$$

where $Z = e + fz^n$; and, since the integrals for $\theta = 1$ (Form I) and $\theta = 0$ (Form II) have been calculated from first principles, the tables can be completed. However, difficulties arise when $\theta = 0$ in the reduction formula. Saunderson circumvents this by making the transformation $e, f, \eta \rightarrow f, e, -\eta$. This transforms I_θ to $I_{1-\theta}$ and thus I_0 is found from I_1, I_{-1} from I_2, I_{-2} from I_3 and so on. The application of the rule is not always as simple as that; for example, under it Form V

$$\frac{d\dot{z}\,z^{\theta n-1}}{\sqrt{e+fz^n}},$$

becomes

$$\frac{d\dot{z}\,z^{-\theta n+\frac{1}{2}n-1}}{\sqrt{e+fz^n}},$$

which is of Form VI, i.e., I_θ, Form V, becomes $I_{-\theta}$, Form VI. By judicious use of such results, Saunderson was able to show how the eighteen tables were completed.

An earlier commentator on Cotes' work was Henry Pemberton (1694–1771), editor of the third edition of Newton's *Principia* (Cambridge, 1726). Shortly after the publication of *Harmonia Mensurarum*, Pemberton published *Epistola ad Amicum de Cotesii Inventis, Curvarum Ratione Quae cum Circulo & Hyperbola Comparationem Admittunt* (London, 1722). (The *amicus* is James Wilson, Pemberton's friend and literary executor.) Pemberton set out to show the geometrical basis of Cotes' work on integrals, namely, the relationships between the circle and ellipse, and the hyperbola. The work is long and prolix and (perhaps not intentionally) increases one's respect for Cotes' achievement. Pemberton shows how all Cotes' integrals can be derived from the fluxional forms published by Newton in *De Quadratura Curvarum* (London, 1704). Pemberton goes on to give, at great length, the first published proof of Cotes' theorem on the factorisation of $a^n \pm x^n$, the so-called 'Cotes property of the circle'. Pemberton's treatment of Cotes' Form II is interesting. He shows how the two forms can be related (i) e, f, of unlike sign, to the area of a hyperbola as in Cotes' Proposition IV in Logometria, and so to logarithms; and (ii) e, f, of like sign, to the area of an elliptic sector, hence to the area of a circular sector, and so to an angle. All this seems rather backward looking, and E. Montucla suggests (*Histoire des Mathématiques*, Part V, vol. I (4 vols., Paris, 1798), p. 153) that Pemberton was motivated by jealousy of Cotes. There is no clear evidence for this, although it would be reasonable to suppose that Pemberton, keen to be the editor of the third edition of *Principia*, would take the opportunity to demonstrate that he could do what Cotes could.

One further, very detailed and elaborate, commentary on Cotes' work on integrals was written by a young Benedictine monk, Charles Walmesley, born in 1722, the year of the publication of *Harmonia Mensurarum*. Walmesley wrote, usually in French, on a number of mathematical and astronomical topics, and was one of those invited by the British government to work on the reform of the calendar. He is perhaps more widely known for his religious writings under the name of Pastorini. Who better than an English mathematician, born in Wigan, living and working in Paris and writing in French, to give us a balanced mid-eighteenth-century view of Cotes' work? The commentary was published as *Analyse des Mesures* (Paris, 1749). In his Preface, Walmesley noted that it was twenty years since the publication of *Harmonia Mensurarum*, but that the book was not well known. Those few authors who had read it spoke highly of it, but found that Cotes, whilst indicating his discoveries, left his methods wrapped in obscurity. One saw at once that the solutions given to many problems were the simplest yet given by any geometer, and the calculations less laborious than one had been accustomed to by other methods. This new method consisted of reducing integrals which depended on conic sections, to measures of ratios and of angles. All that was needed were tables of logarithms, sines and tangents, and to be able to find the fourth member in a proportion in which one knew the other three. Newton had brought the integral calculus to such a state of perfection, one would not think it possible to add anything new. Cotes had greatly simplified the methods. Not content with this he had opened up a vast field, until then unknown: finding ways of reducing differentials, until then considered irreducible, and evaluating them by measures of ratios and angles, either together or combined. What enabled him to do this was his discovery of the beautiful property of the circle, which gave him the means of reducing complicated differentials, of which the denominators were binomials, to several other more simple differentials. He extended this to denominators with any number of terms but his result here was less general. Jean Bernoulli had treated the same subject (*Opera Omnia*, vol. II (4 vols., Lausanne, 1742), p. cxiv), but his solution, although elegant, was less convenient than Cotes'. A. De Moivre, in *Miscellanea Analytica* (London, 1730), completed the work on the circle and found the means to resolve those cases which had escaped Cotes.

Here, in Walmesley's Preface, is a clear and benign summary of Cotes' achievement in this particular field; but Walmesley's encomium, and the detailed commentary which followed it, failed to establish the methods at all widely. Cotes' undoubted major advance in integration techniques had already been overtaken by the work of the Continental mathematicians with their rapidly developing techniques, superior notation, and growing understanding of the concept of function. In England, Colin Maclaurin (*Treatise of Fluxions* (4 vols., Edinburgh, 1792)) and Thomas Simpson (*The Doctrine and Application of Fluxions* (2 vols., London, 1750)) gave some indication of Cotes' work, but made little impact. British mathematics was now entering that long sleep, from which it would only awaken under the kiss of the Cambridge Analytical Society in the early morning of the following century.

In the Scholium Generale of Logometria, Part I, Cotes gives pleasingly concise solutions to a number of well-known problems but does not reveal his methods. Robert Smith described Cotes as, 'generally sparing of his words and thoughts, that he might not seem tedious to able mathematicians' and, as far as the problems under discussion are concerned, this is an understatement. However, although Cotes has, in his integral notation, a suitable vehicle for expressing the solutions to many of the problems, in analytical form, we must remember that this notation was not published with Logometria, Part I. The solutions are given verbally, in geometrical terms, as was customary at the time, so that Cotes leaps from a bare statement of the problem to a geometrical construction which embodies the solution. Fortunately, we have a near-contemporary commentary from Nicholas Saunderson, Lucasian Professor after Whiston, and a great admirer of Cotes. The problems solved in the Scholium Generale, and presented without proofs, traditionally gave rise to integrals which were evaluated by term-by-term integration of series, or integration by conic areas. The problems are selected, according to Cotes, 'to demonstrate the superiority of the logarithmic method' and Saunderson, in *The Method of Fluxions*, pp. 162–206, has given a clear analysis of them. In the Preface to that work, one of the publishers, probably Whiston, has written: 'What the Doctor has given us upon Mr Cotes' Logometria is particularly valuable, as by his intimate acquaintance with that extraordinary person, he may be presumed to have understood his writings better than anyone at that time living: Dr Smith only excepted.'

The first two problems concern the rectification of a pair of curves related by the evolute–involute property studied by James Gregory and by Huygens (see, for example [1], p. 233) and, nearer Cotes' time, by Varignon. Briefly, the transformation $y = r$, $x = \int r \, d\theta$, transforms the Cartesian equations of parabolas $y^m = ax^n$ and hyperbolas $y^m x^n = a$ (the evolutes) into polar equations of the parabolic and hyperbolic spirals (the involutes). The point of interest to Cotes, and to the present study, is that under this transformation, $\dot{x}^2+\dot{y}^2$ transforms to $\dot{r}^2+(r\dot{\theta})^2$, i.e., the fluxion of arc is the same in both cases and, therefore, the form of integral will be the same in both cases.

The parabola $y^2=ax$ transforms to the archimedean spiral $r=\frac{1}{2}a\theta$. The fluxion of arc in both cases can be expressed as $(\dot{y}/a)\sqrt{a^2+4y^2}$. From Cotes' tables, the form required is Form IV, for $\theta=0$; the details are summarised below:

Fluxion of the arc

$$\frac{\dot{y}}{a}\sqrt{a^2+4y^2}$$

Cotes' fluxional Form IV

$$d\dot{z}\,z^{\theta n+\frac{1}{2}n-1}\sqrt{e+fz^n}$$

d	e	f	θ	η	z
$1/a$	a^2	4	0	2	y

Cotes' form of the integral

$$\frac{z^n}{\eta}\,dP+\frac{ed}{\eta f}R\left|\frac{R+T}{S}\right.$$

Key

$$P=\sqrt{\frac{e+fz^n}{z^n}}$$

$$R=\sqrt{f}$$

$$T=\sqrt{\frac{e+fz^n}{z^n}}$$

$$S=\sqrt{\frac{e}{z^n}}$$

Cotes' integral evaluated

$$\frac{y}{2a}\sqrt{a^2+4y^2}+\frac{a}{4}\left|\frac{2+\sqrt{\dfrac{a^2+4y^2}{y^2}}}{\dfrac{a}{y}}\right.$$

Note the modern form of the integral:

$$\frac{1}{a}\int_0^y\sqrt{a^2+4y^2}\,dy=\frac{y}{2a}\sqrt{a^2+4y^2}+\frac{a}{4}\sin^{-1}\frac{2y}{a}$$

$$=\frac{y}{2a}\sqrt{a^2+4y^2}+\frac{a}{4}\ln\frac{2y+\sqrt{a^2+4y^2}}{a},$$

which agrees.

In the first Problem: *To rectify the archimedean spiral*, with reference to Appendix 1, Fig. 12,

$$PQ = y, \ QU = PQ \text{ and } PU = \dot{x}.$$

The fluxion of the arc is

$$\frac{\dot{y}}{a}\sqrt{a^2 + 4y^2}$$

and the integral is the same as in the previous case. In both cases the geometrical construction of a straight line equal in length to the arc, is, on the face of it, simple. However, the length *LM* can only be constructed by reference to tables, i.e., it is a measured line. This occurs frequently in the constructions, but this element of approximation subtracts nothing from the aesthetic pleasure which the solutions give.

Cotes was particularly pleased with his rectification of the logarithmic curve, the evolute in the next pair, the involute being the reciprocal spiral, $r = a/\theta$, named the reciprocal spiral by Cotes and the name has remained. As has been mentioned earlier, Cotes sent his rectification of the logarithmic curve to his friend, William Jones, as an example of the kind of work he was engaged upon and, 'as a curiosity which may be understood independently of the rest' ([4], Letter cx). This was in reply to Jones' letter of 1 January 1712 (New Style), in which Jones said that he (Cotes) had been chosen a member of the Royal Society, and urged him to publish some of his work in the *Philosophical Transactions of the Royal Society*, adding that Sir Isaac Newton expected it (see [4], Letter clx). It is clear from the letter which accompanied Cotes' rectification of the logarithmic curve that, although he was willing to send his unpublished pieces to Jones for interest and in friendship, he was much more keen to publish them himself at Cambridge, rather than in the *Transactions of the Royal Society*, and was in fact making preparations to do so.

The equation $x = -a \ln(y/x)$ transforms under $y = r$, $x = \int r \, d\theta$, to Cotes' reciprocal spiral $r = a/\theta$. The fluxion of arc is $(\dot{y}/y)\sqrt{a^2 + 4y^2}$. The constant sub-tangent length is an obvious quantity to choose as the modulus *a*, equal to the constant polar sub-tangent length in the reciprocal spiral. Form iii for $\theta = 0$ is the relevant form, and the details are again summarised below:

In the second Problem: *To rectify the reciprocal spiral, a curve whose polar sub-tangent is constant,*

$$AE = r, \quad \text{angle } BAE = \theta, \quad y = r, \quad \dot{x} = r\dot{\theta}$$

(see also Appendix 1, Fig. 13).

Fluxion of arc

$$\frac{\dot y}{y}\sqrt{a^2+4y^2}$$

Cotes' form of the integral

$$\frac{2}{\eta}\,dP-\frac{2}{\eta}\,d\,R\left|\frac{R+T}{S}\right.$$

Cotes' integral evaluated

$$\sqrt{a^2+y^2}-a\left|\frac{a+\sqrt{a^2+y^2}}{y}\right.$$

Cotes' fluxional Form III

$$d\dot z\, z^{\theta\eta-1}\sqrt{e+fz}$$

d	e	f	θ	η	z
1	a^2	1	0	2	y

Key

$$P=\sqrt{e+fz^{\eta}}$$

$$R=\sqrt{e}$$

$$T=\sqrt{e+fz^{\eta}}$$

$$S=\sqrt{fz^{\eta}}$$

Note the modern form of the integral: taking the equation as $r=a/\theta$, the arc length is

$$\int\sqrt{1+\frac{a^2}{r^2}}\,dr=\sqrt{a^2+r^2}-a\,\ln\left(\frac{a+\sqrt{a^2+r^2}}{r}\right),$$

taken between suitable limits of r. The first term is the length of *EF*, the second term is the measure of $(EF+AF)/AE$ to modulus a. Cotes' construction therefore follows. Notice that $AF/(EF-AF)=(EF+AF)/AE$.

In the third Problem: *To rectify the logistic line (logarithmic curve),* see Appendix 1, Fig. 14, the lettering of the diagram reflects the involute – evolute relationship of the previous curve and this one. With $AF=a$ the constant sub-tangent; x axis along AF; y axis along AE, we have

$$\frac{\dot y}{\dot x}=\frac{y}{a},\qquad\text{i.e., }\dot x=a\frac{\dot y}{y}$$

and the integration to be performed is therefore identical with that in the previous problem.

Note: From the nature of the curve, *LM is* the measure of the ratio of *AE* to *AL*, which ratio, by construction, is equal to *AE* to $(EF-AF)$ as before. Similarly for *lm.*

Enough has been said to illustrate the way in which the tables of integrals could be used to give rapid solutions to problems of contem-

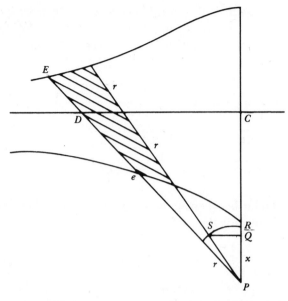

Fig. 7. The conchoid is the locus of points E and e, such that $DE = De = r \cdot P$ is a fixed point, $PC = a$.

porary interest, the commonly accepted touchstone for new methods. Much obviously depends on ingenious use of coordinate systems; for example, in the quadrature of the conchoid, with $PQ = x$, the fluxion of area is

$$\frac{2ar^2\dot{x}}{x\sqrt{r^2-x^2}},$$

which is Form v for $\theta = 0$, where $PR = r$ (Fig. 7). The group of problems on the hyperbola require, variously, Forms III and IV and, as do most of the problems in the Scholium Generale, yield logarithmic solutions. The surface area of the ellipsoid is historically of much greater interest, for it is in connection with this problem that we find Cotes' anticipation of De Moivre's theorem. It arose when considering the two aspects of the problem, logarithmic or trigonometric, according as the spheroid is oblate or prolate. (De Moivre's result was not published until 1731, in *Miscellanea Analytica*.)

Rotation of an elliptic arc about the minor axis yields a surface whose fluxion is of Form IV and has a logarithmic fluent; whilst rotation about the major axis yields a surface whose fluxion is of the same form, but since the modulus involves the term $\sqrt{a^2-b^2}$ in the first case, and $\sqrt{b^2-a^2}$ in the second case (where a and b are the lengths of the semi-axes), the modulus in the second case is complex and the fluent has a trigonometrical form. If the equation of the ellipse is taken in the standard form $(x/a)^2 + (y/b)^2 = 1$, Cotes' construction in the first case is equivalent to

$$\pi a \left\{ \frac{y}{b^2}\sqrt{y^2(a^2-b^2)+b^4} + \frac{b^2}{\sqrt{a^2-b^2}} \ln \frac{\sqrt{y^2(a^2-b^2)+b^4} + y(a^2-b^2)}{b^2} \right\}$$

and, in the second case, is equivalent to

$$\pi a \left\{ \frac{y}{b^2}\sqrt{b^4-(b^2-a^2)y^2} + \frac{b^2}{\sqrt{b^2-a^2}} \cdot X\hat{E}C \right\}.$$

These results, taken between limits 0 and b, and doubled, give the standard expressions for the surface areas,

$$2\pi a^2 + \pi\frac{b^2}{e} \ln \frac{1+e}{1-e} \quad \text{and} \quad 2\pi\frac{ab}{e}\arcsin e + a^2.$$

The latter surface, says Cotes, could have been described by logarithms (see Appendix 1, p. 170), but by an impractical method (*sed modo inexplicabili*). Cotes continues:

Nam si quadrantis circuli quilibet arcus, radio *CE* descriptas, sinum habeat *CX* sinumque complimenti ad quadrantem *XE*: sumendo radium *CE* pro modulo, arcus erit rationis inter *EX* + *XC*√−1 & *CE*, mensura ducta in √−1 (*Harmonia* Mensurum, p. 28).

For if some arc of a quadrant of a circle described with radius *CE* has sine *CX*, and sine of the complement of the quadrant *XE*, taking radius *CE* as modulus, the arc will be the measure of the ratio between *EX* + *XC*√−1 and *CE*, the measure having been multiplied by √−1 [Fig. 8].

In modern notation, this is

$$\text{iR} \ln (\cos \theta + \text{i} \sin \theta) = \text{R}\theta,$$

and we must remain uncertain whether this is anything more than a slip on Cotes' part.

Cubic equations (Appendix 1, p. 171)
In the solution of cubic equations Cotes finds, once again, a system whose different aspects concern either ratios or angles, for example the trigonometrical method of solution gives the three real roots in precisely those cases where the Cardan method fails. He had read Newton's *Algebra* (later published as *Arithmetica Universalis*, ed. W. Whiston (Cambridge, 1707)) as an undergraduate and had made his own manuscript copy. (This is preserved in the Trinity College Library

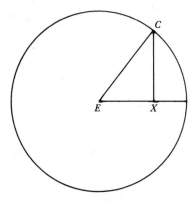

Fig. 8

MS R 16 39, signed 'Roger Cotes, Anno 1700'.) Cotes wrote fairly extensive annotations on *Algebra*, as part of a scheme in which he hoped to get Newton to publish an improved edition. In particular, he devoted some attention to the section dealing with various so-called verging methods – adaptations of straight-edge and compass construc-tions for obtaining approximate solutions to problems whose algebraic solutions usually lead to cubic equations. Cotes' paper on this is among the Clare papers, and is reproduced in Fig. 9.

The gravitational attraction of spheroids (Appendix 1, p. 172)
This very condensed section conceals a great deal of work, most of it arising in connection with *Principia*. Cotes' Form xvi applies to the problems and, by its use, the attraction due to an oblate spheroid at a point on its major axis yields a logarithmic solution, and that due to a prolate spheroid at a point on the minor axis gives a trigonometric solution. However, in Cotes' paper on this subject, prepared for *Principia* (see Trinity College Library MS R 16 38) Cotes uses integra-tion by conic areas, and this is how the results appear in all editions of *Principia* (Book iii, Problem xix, first edition p. 422; second edition p. 380). The attraction due to a spheroid at a general point was a much discussed question in the eighteenth century and I. Todhunter has given a very full account in *A History of the Mathematical Theories of Attraction* (London, 1873).

Spiral orbits under the inverse cube law (Appendix 1, p. 173)
The work here is a good example of Cotes' ability to unify apparently diverse results. Form v is the fluxional form applicable to the group. Cotes makes good use of Newtonian polar coordinates (as given in

Copy of a Paper of Mr Cotes. Communicated to me by Mr Smith Plumian Professor of Astronomy.

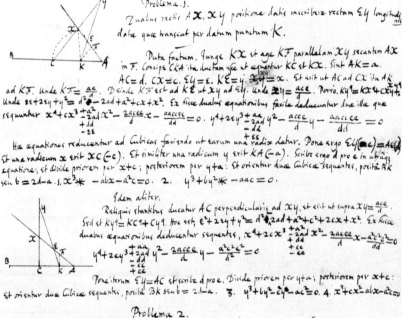

Problema 1.

Dualus rectis AX, XY positione datis inscribere rectam EY longitudinis datæ quæ transeat per datum punctum K.

Puta factum. Junge KX et age KF parallelam XY secantem AX in F. Concipe CKA ita ductam esse ut æquentur KC et KX. Sint $AK = a$.
$AC = d.$ $CX = c.$ $EY = e.$ $KE = y.$ Et erit ut AC ad CX ita AK ad $KF.$ Unde $KF = \frac{ac}{d}.$ Deinde KF est ad KE ut XY ad $EY.$ Unde $XY = \frac{ace}{dy}.$ Porro. $KY = KX + CXY$
Unde $aa + 2ay + y^2 = d^2 - 2ad + a^2 + cx + x^2.$ Ex hisce duabus æquationibus facile deducuntur duæ illæ quæ

sequuntur $x^4 + cx^3 + \frac{+aa}{-2ad}x^2 - \frac{2acce}{dd}x - \frac{aacce}{dd} = 0.$ $y^4 + 2cy^3 + \frac{+aa}{-2ad}y^2 - \frac{acce}{d}y - \frac{aaccee}{dd} = 0$

Hæ æquationes reducentur ad Cubicas faciendo ut earum una radix detur. Pone ergo $EY(=e) = AC$
Et una radicum x erit $XC(=c).$ Et similiter una radicum y erit $KA(=a).$ Scribe ergo d pro e in utraque
æquatione; et divide priorem per $x + c$; posteriorem per $y + a.$ Et orientur duæ Cubicæ sequentes, positâ BK
sub $b = 2d - a.$ 1. $x^3 * - abx - a^2c = 0.$ 2. $y^3 + by^2 * - aac = 0.$

Idem aliter.

Reliquis stantibus ducatur AC perpendicularis ad $XY,$ et erit ut supra $XY = \frac{ace}{d}.$
Sed et $KY = KC + CY.$ Hoc est, $e^2 + 2ey + y^2 = d^2 + 2ad + a^2 + c^2 + 2cx + x^2.$ Ex hisce
duabus æquationibus deducentur sequentes, $x^4 + 2cx^3 + \frac{+aa}{-2ad}x^2 - \frac{2acce}{d}x - \frac{a^2c^2}{dd} = 0$

$y^4 + 2cy^3 + \frac{+aa}{+2ad}y^2 - \frac{2acce}{d}y - \frac{a^2c^2}{d^2} = 0$

Pone iterum $EY = AC$ et scribe d pro $e.$ Divide priorem per $y + a$; posteriorem per $x + c$:
Et orientur duæ Cubicæ sequentes, positâ BK sub $b = 2d - a.$ 3. $y^3 + by^2 - a^2 = 0.$ 4. $x^3 + cx^2 - abx - a^2c = 0$

Problema 2.

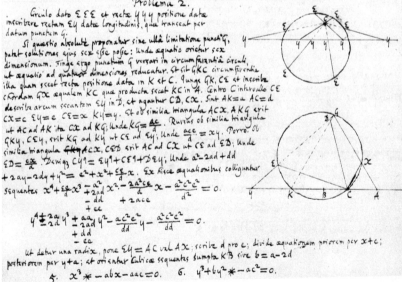

Circulo dato EEE et rectæ YY positione datæ
inscribere rectam EY datæ longitudinis, quæ transeat per
datum punctum $G.$

Si quæstio absolutè proponatur sine ullâ limitatione puncti $G,$
patet solutiones ejus sex esse posse: Unde æquatio oritur sex
dimensionum. Finge ergo punctum G versari in circumferentiâ circuli,
ut æquatio ad quatuor dimensiones reducatur. Et sit GKC circumferentia
illa quam secet recta positione data in K et $C.$ Junge $GX,$ CE et inscribe
chordam GX æqualem KC quæ producta secat KC in $A.$ Centro C intervallo CE
describe arcum secantem EY in $D,$ et agantur $CD, CX.$ Sint $AK = a$ $AC = d$
$CX = c$ $EY = e$ $CE = x$ $KY = y.$ Et ob similia triangula ACX, AKG erit
ut AC ad AK ita CX ad $KG;$ Unde $KG = \frac{ac}{d}.$ Rursus ob similia triangula
$GKY, CEY;$ erit KG ad KY ut CE ad $EY;$ Unde $\frac{ace}{d} = xy.$ Porro. Ob
similia triangula $GACX, CED$ erit AC ad CX ut CE ad $ED;$ Unde
$ED = \frac{cx}{d}.$ Deinde $CY^2 = EY^2 + CE^2 + 2D \cdot EY;$ Unde $a^2 - 2ad + dd$
$+ 2ay - 2dy + y^2 = e^2 + x^2 + \frac{2cx}{d}x.$ Ex hisce æquationibus colliguntur

sequentes $x^4 + \frac{cx}{d}x^3 + \frac{+aa}{+2ad}x^2 - \frac{2a^2ce}{dd}x - \frac{a^2c^2e^2}{d^2} = 0.$
$ \frac{-dd}{+ee} + 2ace$

$y^4 \frac{+2ay}{-2dy}y^3 + \frac{+aa}{-2ad}y^2 - \frac{ace^2}{dd}y - \frac{a^2c^2e^2}{dd} = 0.$
$ \frac{+dd}{-ee}$

Ut detur una radix, pone $EY = AC$ vel $AX;$ scribe d pro $e;$ divide æquationem priorem per $x + c;$
posteriorem per $y + a;$ et orientur Cubicæ sequentes sumptâ KB sive $b = a - 2d$

5. $x^3 * - abx - aac = 0.$ 6. $y^3 + by^2 * - ac^2 = 0.$

Fig. 9. Cotes' paper on verging problems. (By permission of Clare College
Library, Cambridge.)

Methodus Fluxionum, seventh method). Details for Case 1 will serve as an illustration. The diagrams for the five spiral orbits are shown in Appendix 1, Figs. 26–30. With reference to Fig. 26, in which I have added the points p and π to Cotes' diagram, P and p are neighbouring points on the curve. SP meets the construction circle in R. π is a point on SP such that $S\pi = Sp$. (The notation is Saunderson's. His analysis is equivalent to $\dot{r}/r\dot{\theta} = \tan \phi$ in modern conventional notation.)

$$\frac{\pi P}{p\pi} = \frac{PQ}{QS} \quad \text{or} \quad \frac{\dot{x}}{x\dot{\theta}} = \frac{\sqrt{x^2 - QS^2}}{QS}. \tag{1}$$

We need two results from *Principia*:

(i)   Under the given initial conditions, the velocity will be proportional to $\sqrt{a^2 - x^2}/x$ (Book I, Section VII, Proposition XXXIX).

(ii) QS is proportional to $1/v$ (Book I, Section II, Proposition I), $= bx/\sqrt{a^2 - x^2}$.

Putting $n^2 = a^2 - b^2$, and taking n as the radius of the circle, (1) reduces to

$$n\dot{\theta} = \frac{-bn\dot{x}}{x\sqrt{n^2 - x^2}} \quad (x \text{ decreasing as } \theta \text{ increases}),$$

a fluxion of Form V, for $\theta = 0$. The fluent, taken between 0 and θ for θ, n and x for x, is

$$n\theta = b \left| \frac{n + \sqrt{n^2 - x^2}}{x} \right.$$

DSE (Fig. 10) is clearly the RTS triangle, and Cotes' explanation of the construction is clear. The modulus is $SM = b$ (Appendix 1, Fig. 26), and there is a one-to-one correspondence between points D on SA and points of the orbit. By measuring the arcs $n\theta$ on the other side of A, the trajectory corresponding with projection in the opposite direction is constructed. The orbit is clearly symmetrical about SA, x cannot exceed n and is only equal to n when θ is 0. Therefore, A is the apse. There is, however, no upper or lower bound to the values of θ, hence Cotes' final observations.

Before considering the velocity and the time, it is of interest to note that

$$n\theta = b \ln \left\{ \frac{n + \sqrt{n^2 - x^2}}{x} \right\}$$

can be written as

$$u = \tfrac{1}{2}n\{e^{n\theta/b} + e^{-n\theta/b}\},$$

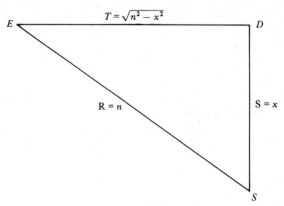

Fig. 10. The triangle *DSE* is abstracted from Appendix 1, Fig. 26. From *Harmonia Mensurarum*, Cambridge University Press, 1722.

where $u = 1/x$. This is a solution of $d^2u/d\theta^2 = (n^2/b^2)$. In Robert Smith's own copy of *Harmonia Mensurarum* (Trinity College Library MS adv b 115), the words 'Hyperbolic Spiral', 'Logarithmic Spiral', 'Anonymous Spiral', 'Reciprocal Spiral', 'Elliptic Spiral', have been written against the respective cases.

Newton's investigation of Cases 1 and 5 are in *Principia*, Book I, Section VIII, Proposition XLI, and the integration is of course by conic areas. Cotes seems never to have told Newton about his integration techniques, but his comment in Logometria, Part I about 'some theorems, not only of logometry, but also of trigonometry, which I keep ready for use', suggests that he made full use of the theorems in the course of his editorial work. These theorems are the tables of integrals already discussed, and which appeared in *Harmonia Mensurarum*, under the title 'Theoremata tum Logometrica tum Trigonometrica'.

The motion of a cycloidal pendulum in a resisting medium (Appendix 1, p. 179)

Among Cotes' Opera Miscellanea, which constitute the latter part of *Harmonia Mensurarum*, is a tract called 'De Motu Pendulorum in Cycloide'. This is one of Cotes' early papers, probably intended for lectures to undergraduates. There is a manuscript copy dated 1708 in the Clare papers. In this tract, the motion of a cycloidal pendulum in a non-resisting medium is fully set out, and the isochronous property, well known since Huygens' day, established. Cotes made a brief reference to this paper in a letter to his uncle: 'I thank You for ye kind Judgement You made concerning my Paper about Projectiles. I have by me another such a Paper concerning ye motion of Pendulums which I drew up about ye same time with y^{tt} (Cotes to John Smith, 16 October 1707 [4], Letter XCVIII).

The much more general consideration of resisted cycloidal motion is Proposition XXIX, Book II, in *Principia*. Cotes is again concerned to present a more analytical solution, but now without reference to the tables of integrals. His use of simple harmonic motion in a straight line as a model of unresisted cycloidal motion has a modern flavour, and his solution of the differential equation for resisted motion is ingenious. The equation obtained by Cotes is, in modern form,

$$u\frac{du}{dx} - ku^2 + x = 0,$$

a Bernoulli-type equation, proposed for solution by Jacques Bernoulli in the *Acta Eruditorum* (1695), p. 533, and solved by Leibniz in the *Acta Eruditorum* (1696), p. 145.

The penultimate Problem in this long Scholium Generale is also a *Principia* problem, derived from Proposition XXII, Book II. It is: *To find the density of the atmosphere at any distance from the centre of the earth, gravitational attraction varying as any power whatever of that distance.* In this, as in the preceding, and succeeding (final) problems, the steps of the argument are clearly set out in a deductive manner, and the solutions obtained without the use of the tables of fluents. It is as though Cotes sought to forestall any suggestion that his solutions were obtained by the use of some sort of bag of tricks. The final Problem (Appendix 1, p. 184) is introduced with the words: 'Finally, so that it may be more fully established that synthetic demonstrations may be devised with little trouble from the above principles, it will suffice to add just one more example' – a statement which would make the methods acceptable to Cotes' more traditionally-minded contemporaries.

And so to the final Problem: *The calculation of the length of the meridian corresponding to a given latitude range, in the stereographic projection.* The argument brings us right back to Proposition I of Logometria, thus neatly rounding off this very substantial, first, and only publication.

LOGOMETRIA, PART III

Here Cotes presents a delightful collection of twelve problems, demonstrating not only the range and effectiveness of the methods, but also the details of the proofs. In a brilliant display of ingenious coordinate systems, allied to integration techniques, he develops a geometrical analysis to which traditionally-difficult problems yielded easily.

Abandoning his former cautious concealment of proofs, he sets out
the details in an expansive style clearly intended for a wider audience.
We find a further range of problems on arc lengths, surfaces and
volumes of revolution, and areas. John Wallis had rectified the cissoid
in *Tractatus Duo* (Oxford, 1695). Huygens had also rectified the cissoid
as well as the parabola, and found the parabolic surface of revolution.
In the *Acta Eruditorum* (of 1691), Jacques Bernoulli had related the
parabola to the spiral of Archimedes through the transformation $r = y$,
$x = \int r \, d\theta$ (previously used by James Gregory). Somewhat earlier, Torricelli had investigated the properties of the logarithmic curve, finding
the area between the curve and its asymptote and the corresponding
volume of the solid of revolution; and also the finite surface generated
by an infinite arc of a hyperbola. Cotes drew on all these results to
demonstrate the superiority of his techniques. Precise identification
of his actual sources remains uncertain, except in the case of the
problems on central orbits, gravitational potential and resisted motion;
these are a direct development of his work on *Principia*. There is a
further, entirely analytical consideration of the problem of the Mercator distances; and as a final flourish, the rectification and construction of the catenary, a clear reference again to the work of Jean
Bernoulli in *Acta Eruditorum* (1691).

The twelve problems selected by Cotes are

I   *The quadrature of the tangent curve.*

II   *The quadrature of the secant curve.*

III   *The volume of the solid of revolution formed from the tangent curve.*

IV   *The properties of the enclosed, or spiral, tractrix, and of its negative
pedal* (named by Cotes, the *lituus*).

V   *The surface area of the paraboloid of revolution.*

VI   *The surface areas of cissoidal solids of revolution.*

VII   *The surface area generated by revolution of the logarithmic curve
about its asymptote.* Scholium on finite and infinite surfaces of
revolution.

VIII   *The angular velocity of a sphere and its circumscribing cylinder
about their common axis.*

IX   *The attraction due to a sphere, under the inverse cube law.*

X   *Motion under a centripetal force varying as the inverse cube of the
distance* (the Cotes spiral).

XI   *The vertical motion of a heavy body in a resisting medium.*

XII   *To construct a catenary.*

In Problems I: *The quadrature of the tangent curve*; and II: *The
quadrature of the secant curve*, Cotes is really concerned with exploiting

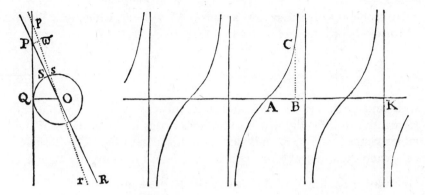

Fig. 11. The tangent curve as it appeared in *Harmonia Mensurarum*, Cambridge University Press, 1722.

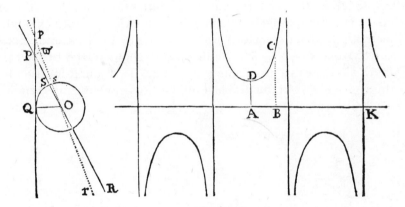

Fig. 12. The secant curve as it appeared in *Harmonia Mensurarum*, Cambridge University Press, 1722.

the Mercator distances problem which involves the integral of sec θ. The diagrams are shown in Figs. 11 and 12 as they appeared in *Harmonia Mensurarum*, the first time these curves had appeared in print. With a generating circle of radius r, and $AB = r\theta = x$, $BC = QP = y$, the fluxion of area in Problem I is $(r^2 y\dot{y})/(r^2 + y^2)$, which is Form I for $\theta = 1$ and the fluent is

$$\tfrac{1}{2}r^2 \left| \frac{r^2 + y^2}{r^2} \right. ,$$

equivalent to $r^2 \ln \sec \theta$.

In Problem II the fluxion of area is $(r^2\dot{y})/\sqrt{y^2-r^2}$, which is of Form VI for $\theta = 0$ and the tables give the fluent as

$$\frac{2}{\eta f}\,d\mathrm{R}\left|\frac{\mathrm{R}+\mathrm{T}}{\mathrm{S}}\right.,$$

with suitable values for R, T and S. In this case $S = \sqrt{-r^2/y^2}$, i.e., S^2 is negative, a difficulty which sometimes arises in applications of the tables. Since $\mathrm{S}^2 = \mathrm{T}^2 - \mathrm{R}^2$, it is clear that this arises when R^2 is positive, i.e., when the integral represents an angle, and R^2 is greater than T^2. One way of dealing with this, is to write $\mathrm{R}\left|\dfrac{\mathrm{R}+\mathrm{T}}{\sqrt{-\mathrm{S}^2}}\right.$ as

$$\mathrm{R}\left|\frac{\mathrm{R}+\mathrm{T}}{\mathrm{S}}\right. + \mathrm{R}\left|\frac{1}{\sqrt{-1}}\right.,$$

and to ignore the last term which, as Smith says, is at least a constant (and so can be ignored in a definite integral). Cotes advocates another method, which is to replace T with T′, where $\mathrm{R}/\mathrm{T} = \mathrm{T}'/\mathrm{R}$, and recalculate S′ from $\mathrm{S}'^2 = \mathrm{T}'^2 - \mathrm{R}^2$. A little algebra shows this to be the same thing. Cotes gives a list of alternative values T′, S′ and R for use when this is necessary. In the problem under discussion, the integration of sec θ, Cotes gives

$$\mathrm{R}:\mathrm{T}:\mathrm{S}::y:\sqrt{y^2-r^2}:\sqrt{-r^2}$$

and

$$\mathrm{R}:\mathrm{T}':\mathrm{S}'::y:\sqrt{y^2-r^2}:r.$$

Using the latter ratios, and evaluating between suitable limits, the result is

$$OQ^2\left|\frac{OP+PQ}{OQ}\right.,$$

equivalent to $r^2 \ln (\sec \theta + \tan \theta)$.

In the Scholium to Problem II, Cotes returns to the problem of the spacing of the parallels on the Mercator chart. On this chart, all parallels of latitude are made equal in length to the equator, i.e., each is increased in length by a factor sec ϕ, where ϕ is the latitude angle. Therefore, to preserve the orthomorphic properties of the chart, the meridional scale must be stretched by a factor sec ϕ at every point. Hence, the spacing of the parallel from the equator must be $\int_0^\phi r \sec \phi \, d\phi$, r being the radius of the earth. Cotes gives the result equivalent to $r \ln (\sec \phi + \tan \phi)$ and an alternative form, equivalent to

$$r \ln \tan \left(\frac{\pi}{4} + \frac{\phi}{2} \right).$$

Problem III: *To find the volume of the solid of revolution formed from the tangent curve*, is straightforward enough, requiring Form II for $\theta = 1$ to evaluate the integral of $(y^2 \dot{y})/(r^2 + y^2)$.

Problem IV: *To construct a curve such that the arc length intercepted between two bounding radii should be the measure of the ratio of those radii*, is of rather more interest. In 1704, Paul Varignon ('Nouvelle Formation de Spirales', *Histoire et Mémoires de l'Academie, Paris* (April, 1704), pp. 69–131) had discussed the logarithmic properties of spirals. He observed that in a spiral we are concerned with three elements; arc lengths, radii and angles. If, therefore, one of these elements varies in an arithmetic progression, whilst some other varies in a geometric progression, the spiral will have logarithmic properties. Varignon concluded that there were therefore six types of logarithmic spirals, the third of which was the spiral tractrix, examined here by Cotes.

A property of the curve is the constancy of the length of the polar sub-tangent, and this offers itself as a modulus. Using Newtonian polar coordinates, Cotes obtains the differential equation

$$\dot{x} = \frac{m \dot{y} \sqrt{m^2 - y^2}}{y},$$

where $y = r$, $x = r\dot{\theta}$, in conventional modern notation. The length of the polar sub-tangent is m. Form IV for $\theta = 1$ gives

$$x = \frac{m}{y} \sqrt{m^2 - y^2} - m \left| \frac{R+T}{S} \right.,$$

where $R:T:S :: y : \sqrt{m^2 - y^2} : m$ (or $\theta = [(m\sqrt{m^2 - r^2})/r] - \arccos (r/m)$, the conventional modern polar equation of the curve. There follows an attractive analysis of the geometrical properties of the curve, which enables the curve to be constructed (see Fig. 13). Note in passing that points such as C, E are inverse with respect to the generating circle, but it was not until the mid-nineteenth century that Möbius gave formal recognition to this sort of relationship.

The differential equation of the tractrix itself had been given by Jean Bernoulli in the *Acta Eruditorum* for 1691 as

$$\frac{dy}{ds} = \frac{y}{a} \quad \text{or} \quad \frac{dy}{dx} = \frac{y}{\sqrt{a^2 - y^2}}.$$

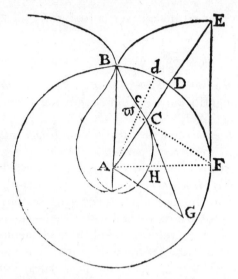

Fig. 13. The spiral tractrix, in which the polar sub-tangent *AG* is of constant length. From *Harmonia Mensurarum*, Cambridge University Press, 1722.

In this problem Cotes brings together ideas from Varignon, Bernoulli and James Gregory. The spiral tractrix is the involute, in James Gregory's sense, of the tractrix. The substitution $y = r$, $\dot{x} = r\dot{\theta}$, in Bernoulli's differential equation above, gives

$$\theta = r^4 + \frac{r^2}{\left(\dfrac{dr}{d\theta}\right)^2},$$

the differential equation of the spiral tractrix in modern form. In the tractrix itself, the arc length is proportional to the logarithm of the bounding ordinates, and this is the property which is carried over into the spiral. Continuing this train of investigation, Cotes adds a Scholium: *To construct the curve such that the area intercepted between two radii shall be the measure of the ratio of the bounding radii.* This is clearly analogous to the logarithmic property of the hyperbola, and applying the transformation $y = r$, $x = r\theta$ to the curve $xy = c$, we obtain $r^2 = c/2\theta$. Cotes obtains $r^2 = a^2/\theta$ the equation of a hyperbolic concentric spiral. The curve is shown in Fig. 14. It was named by Cotes *lituus* (from the Roman augural staff), which name, like reciprocal spiral, has passed into the literature. He adds the comment that this curve is related to the hyperbola in the same way that the archimedean spiral is related to the parabola, and the reciprocal spiral to the logarithmic curve.

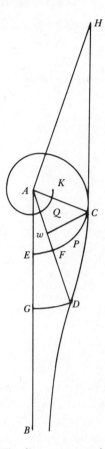

Fig. 14. The *lituus*, so named by Cotes because of its resemblance to the Roman augers' crooked staff. From *Harmonia Mensurarum*, Cambridge University Press, 1722.

Problems v: *The surface area of the paraboloid of revolution*; vi: *The surface areas of cissoidal solids of revolution*; and vii: *The surface area generated by revolution of the logarithmic curve about its asymptote*, concern surfaces of solids of revolution derived from the parabola and (again) from the cissoid and the logarithmic curve. (Huygens had studied the parabolic surface of revolution.) Cotes rotates the parabola about a line perpendicular to its axis. The integration is quite difficult, requiring Form vi for $\theta = 0$ and for $\theta = 1$.

The cissoidal surface of revolution about the axis presents a rather more difficult problem, enabling Cotes to demonstrate extensions of his technique. The considerable interest attaching to the cissoid in the seventeenth century stemmed partly from its well-known use as an

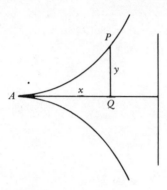

Fig. 15

auxiliary curve in the solution of the Delian problem. With $AQ = x$, $QP = y$ (Fig. 15) and the equation of the curve as

$$y = x\sqrt{\frac{x}{a-x}}.$$

Cotes obtains the fluxion of arc as

$$\frac{a\dot{x}x^{\frac{3}{2}}\sqrt{4a-3x}}{(a-x^2)}.$$

This is not, as it stands, one of the eighteen forms. However, by writing $x^{\frac{3}{2}}$ as $x^{\frac{5}{2}-1}$ it can be expressed as a combination of Form VIII for $\theta = 0$ and for $\theta = -1$ by the use of the second of Cotes' continuation formulae (see Appendix 3). The working is long and complicated, requiring something of a tour de force. Not unreasonably, Cotes omits it, giving the result only, and summarising it in a neat geometrical construction.

In the Scholium to this problem: *The surface area generated by rotating the cissoid about its asymptote* (Form XI for $\theta = 1$); and in Problem VII: *The surface area generated by rotating the logarithmic curve about its asymptote* (Form II for $\theta = 0$), we are concerned with areas generated by infinite arcs. Finite areas and volumes generated by infinite arcs had given rise to much discussion about the nature of indivisibles in the seventeenth century and indeed earlier. (See, for example, Torricelli, 'The Volume of an Infinite Solid', summarised with commentary, in [9], pp. 227–31.) Cotes adds a note:

In this problem [Problem VII] and in the preceding scholium, we have given examples of finite surfaces generated by infinite arcs. Now when the curvilinear area between an infinite arc and its asymptote is finite, the surface area generated by the rotation of the arc about the asymptote is necessarily

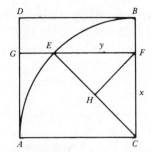

Fig. 16. Rotation of this figure about *BC* generates a hemisphere and its circumscribing cylinder. From *Harmonia Mensurarum*, Cambridge University Press, 1722.

finite. On the other hand, when the area is infinite, the surface will necessarily be infinite.

There is no further discussion, nor proof offered. In a further Scholium: *The surface area generated by rotating a hyperbola about its axis*, is shown to be infinite (Form IV $\theta = -1$). Nicholas Saunderson solves the same problem (see [8], pp. 226–30) by using Form III and adds a note: 'When an asymptotic surface is finite, when infinite', and demonstrates Cotes' conclusion to be true.

In Problems VIII, IX and X, Cotes returns to three problems arising from *Principia*, some aspects of which had been dealt with in Logometria, Part I. Problem VIII is of interest in that Cotes does not seem to have a clear idea of angular momentum although, even in his undergraduate days, he was familiar with the idea of the centre of oscillation of a compound pendulum, probably from reading Huygens' work (see [4], Letter XCVII, p. 197).

In Problem VIII we find a rather curious result. The problem is: *Invenire motum Sphaerae & circumscripti Cylindri, dum simul revolvuntur circum axem utrique communem (To find the quantity of motion of a sphere and its circumscribing cylinder, which rotate simultaneously about their common axis)*. With reference to Fig. 16, the method adopted is to consider the segment of the sphere generated by the rotation of the area *AEFC*, and the circumscribing cylinder of this segment generated by the rotation of the area *AGFC*, both about *CB*. Cotes says the quantities of motion of these two parts are as the fluxions of $\dot{x}y^3$ and $\dot{x}r^3$. Clearly, he is using elementary volumes proportional to $y^2\dot{x}$ and $r^2\dot{x}$, and a wrong result about angular momentum, assuming Cotes is using the term motus in the Newtonian–Cartesian sense of quantity of motion. Of course, y^3 and r^3 should be replaced by y^4 and r^4. Using Form IV Cotes arrives at a ratio for the angular momenta of the two bodies, expressed in geometrical form, equivalent to $3\pi/16$, instead of the correct result 8/15. The term 'moment of inertia' had not then been coined; Euler, who was to introduce it ('Recherches sur la Connoissance Mécanique des Corps, *Histoire de l'Academie Royale des Sciences et Belles Lettres*, 14 (Berlin, 1758) 131–93) being a child of only five years in 1712.

Principia Proposition XCI, Book I, is: *To find the attraction of a corpuscle situated in the axis of a spheroid, the force of attraction decreasing in any ratio of the distance whatever.* In his Problem IX: *The attraction due to a sphere, under the inverse cube law,* Cotes discusses the case of the sphere, the attraction being as the inverse cube of the distance. The use of Form I and some neat geometry gives a pleasing solution which embraces the cases of the particle inside or outside the sphere. Both in *Principia*, and in Logometria, Part I, Cotes gave a lot of attention to the attraction of spheroids, although in his work for *Principia*, as far as his surviving papers can tell us, he kept to integration by conic areas. However, in view of his own comment in the Preface to Logometria, Part I (previously quoted), that he kept certain logarithmic and trigonometric theorems by him 'ready for use', it is likely that Cotes used his integration techniques to develop results, before framing them in the requisite form for *Principia*.

Problem X: *Motion under a centripetal force varying as the inverse cube of the distance*, is a further demonstration of an orbit obtained under the inverse cube law, this time the force being one of repulsion. The resulting orbit is now known as the Cotes spiral, and occurs in the study of potential theory, particularly in its application to hydrodynamics. It is the path of a vortex which moves in a liquid confined between two vertical planes inclined to each other at an angle of $2\pi/n$. Its equation is then $r \cos n\theta = c$, n and c being constants (see A. G. Greenhill, 'Plane Vortex Motion', *Quarterly Journal of Pure and Applied Mathematics*, 15 (London, 1878), 12–15). The analysis and geometrical construction are of sufficient interest to be given in detail; they are therefore included in Appendix 4.

Problem XI: *The vertical motion of a heavy body in a resisting medium*, is a recasting of the problem, the resistance depending on the square of the velocity. Geometrical forms give way entirely to analytical methods, and the spirit is modern. Cotes sets up the differential equations using pricked letters as though they were differentials. He solves the equations by reference to the tables of integrals, obtaining expressions for time, distance and velocity, for both upward and downward projections. It is a nice demonstration that the fluxional notation is reasonably versatile, and perhaps that was the intention.

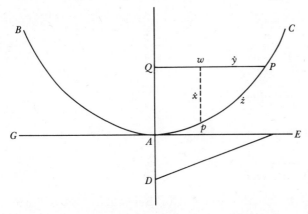

Fig. 17. Cotes rectified the catenary by considering the equilibrium of the element Pp. From *Harmonia Mensurarum*, Cambridge University Press, 1722.

It seems, however, that Cotes (or just possibly his editor Smith) is having doubts about his dual notation $R\left|\dfrac{R+T}{S}\right.$. In some of these later problems, we find $R(R \cdot T/S)$ and even, for example, $R(Rad : tang :: L : V)$, in cases where an angle is intended.

The first transcendental curve to be rectified was the catenary. In response to Jacques Bernoulli's challenge in *Acta Eruditorum* for 1690 – Problem XII: *To find the form of the catenary* – Jean Bernoulli, Leibniz and Huygens submitted solutions, but not methods. David Gregory had also submitted a paper on the catenary for the *Philosophical Transactions of the Royal Society*, subsequently published in *Miscellanea Curiosa*, ed. E. Halley, vol. I (3 vols., London, 1708), p. 129. Gregory's solution was geometrical, making use of a hyperbola and a parabola as auxiliary curves. Cotes, like Jean Bernoulli, considered the equilibrium of part of the catenary, and so derived, as Bernoulli had done, the differential equation of the curve.

With reference to the diagram (Fig. 17), Cotes' solution makes a fitting end to an impressive piece of work. It can be briefly described as follows. P and p are neighbouring points on the curve. Ppw will serve as a triangle of forces for the equilibrium of the arc AP, which is held in equilibrium by the horizontal force from AB, proportional to its length a (a result assumed known); the weight vertically downward of the arc AP proportional to its length z; and the tangential tension at P. Hence,

$$\frac{\dot{x}}{\dot{y}} = \frac{z}{a}$$

and hence,

$$\frac{\dot{x}}{\sqrt{\dot{x}^2 + \dot{y}^2}} = \frac{z}{\sqrt{a^2 + z^2}} \quad \text{or} \quad \dot{x} = \frac{z\dot{z}}{\sqrt{a^2 + z^2}}$$

and

$$z = \sqrt{2ax + x^2},$$

so we have finally

$$\dot{y} = \frac{a\dot{x}}{x^{\frac{1}{2}}\sqrt{2a + x}},$$

the differential equation derived by Bernoulli. This is Form VI for $\theta = 0$ and gives the result

$$y = a \left| \frac{a + x + z}{a} \right.,$$

with suitable attention to limits, or $y + a \ln a = \ln (a + x + 2ax + x^2)$. (This is now conventionally written as $x = a \cosh (y/a) - a$.) Bernoulli, not having Cotes' logarithmic notation in 1691, gave the solution of the differential equation in geometrical terms. Cotes also finishes with a geometrical construction for the catenary, based on his analytical solution of the equation.

4

Cotes' factorisation theorem

The work on Logometria, Parts I, II and III, seems to have been completed before the end of 1712 but, towards the end of his life, Cotes was again working on integrals. William Jones wrote to Cotes on 11 July 1713, thanking him for his copy of the new edition of *Principia* (second edition (Cambridge, 1713)). After complimenting him on it and, in particular, on the Preface, Jones added: 'Now this great Task being over, I hope you'l think of publishing your own Papers, & not let such valuable pieces lye by' (Cotes to Jones, 11 July 1713 [4], Letter XCV).

As we have seen, Logometria, Part I was submitted, rather cautiously, for publication in the *Philosophical Transactions of the Royal Society*, and duly appeared in the issue dated 1714 (vol. 29, no. 338, pp. 5–45). Cotes developed the work further and, on 5 May 1716 (two months before his death), wrote in exuberant tones to Jones:

Geometers have not yet promoted the inverse method of fluxions, by conic areas, or by measures of ratios and angles, so far as it is capable of being promoted by these methods. There is an infinite field still reserved, which it has been my fortune to find an entrance into. Not to keep you any longer in suspense, I have found out a general and beautiful method by measures of ratios and angles for the fluent of any quantity which can come under this form $\dfrac{d\dot{z}\,z^{\theta\eta+\frac{\delta}{\lambda}\eta-1}}{e+fz^{\eta}}$, in which d, e, f are any constant quantities, z the variable, η any index, θ any whole number affirmative or negative, $\dfrac{\delta}{\lambda}$ any fraction whatever. (See S. J. Rigaud, Ed., *Correspondence of Scientific Men of the Seventeenth Century* (3 vols., Oxford, 1841).)

Cotes went on to acknowledge that Newton's first two forms, $(d\dot{z}\,z^{\theta\eta-1})/(e+fz^{\eta})$ and $(d\dot{z}\,z^{\theta\eta+\frac{1}{2}\eta-1})/(e+fz^{\eta})$, were special cases from which the rest of Newton's forms, including the irrational ones, could be deduced.

In the same way the fluents of such forms as

$$\dot{z}\, z^{\theta n-1}(e+fz^{\,n})^{\delta/\lambda};$$

(Cotes–Smith Form VI)

$$\dot{z}\, z^{\theta-1}\left\{\left(\frac{e+fz^{\,n}}{g+hz^{\,n}}\right)\right\}^{\delta/\lambda},$$

(Cotes–Smith Forms IX and X)

see later, could be deduced from his (Cotes') more general forms, and depended on the measures of ratios and angles. A reference to Cotes' eighteen forms will show that nothing more complicated than $\delta/\lambda = 1/2$ occurs. Clearly, Cotes had developed the work substantially further since writing Logometria, Part II. He goes on in the same letter to say that he can

Show by measures of ratios and angles, without any exception or limitation, the fluent of this general quantity,

$$\frac{\dot{z}\, z^{\theta n+\frac{\delta}{\lambda}n-1}}{e+fz+gz^{2n}} \quad \text{or} \quad \frac{\dot{z}\, z^{\theta n+\frac{\delta}{\lambda}n-1}}{e+fz^{\,n}+gz^{2n}+hz^{3n}},$$

where the quantities are as before, except that λ is restricted to be a power of 2.

In a somewhat uncharacteristically caustic passage, Cotes refers to Leibniz' claim that the fluent of $\dot{x}/(x^4+a^4)$ cannot be expressed by measures of ratios and angles. Leibniz had said that the fluent of $\dot{x}/(x+a)$ being a measure of a ratio, and that of $\dot{x}/(x^2+a^2)$ being a measure of an angle, it would be of interest to know to what the fluents of $\dot{x}/(x^4+a^4)$, $\dot{x}/(x^8+a^8)$, and so on, could be referred (see *Acta Eruditorum* (1702), p. 218). 'His desire is answered [said Cotes] in my general method, which contains an infinite number of such progressions.' Cotes concludes by expressing his belief that it will be found that the fluents of all rational fluxions can be expressed as measures of angles or of ratios (i.e., Cotes is suggesting that all polynomials can be expressed as products of linear and/or quadratic factors).

This reference to a general method led Robert Smith to search among Cotes' papers, as he tells us in his Preface to *Harmonia Mensurarum* (Cambridge, 1722). From this Preface we learn that Cotes had indeed resumed work on his tables of integrals early in 1716. In searching among the papers, which appear to have been left in some disorder at the time of Cotes' early and unexpected death, Smith says he 'recovered from the ruin' the *Theoremum Pulcherrimum* which was the basis of the method (Cotes' 'general and beautiful method') but

Fig. 18. Portrait of Robert Smith, by Vanderbank, 1730. The portrait, which hangs in the Master's Lodge at Trinity College, Cambridge, was presented to the College by Thomas Riddell in 1827. (By permission of the Master, Trinity College, Cambridge.)

'neither by line nor by word was it explained'. Smith then presented the theorem, without proof; we may summarise Smith's statement of the theorem as follows.

If the factors of the binomial $a^\lambda \pm x^\lambda$ are required, λ being an integer, the circumference of a circle centre O, radius a is divided into 2λ equal parts, AB, BC, CD, DE, EF and so on [Fig. 19]. *P is joined to A, B, C, D, E, F and so on, where P is on OA and OP = x. Then the product*

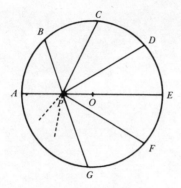

Fig. 19. Cotes' property of the circle leads to the factorisation of $a^\lambda \pm x^\lambda$.

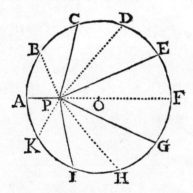

Fig. 20. Robert Smith's diagram for Cotes' property of the circle. From *Harmonia Mensurarum*, Cambridge University Press, 1722.

AP · CP · EP · · · = $a^\lambda - x^\lambda$, or $x^\lambda - a^\lambda$, according as P is inside or outside the circle, and in either case, the product BP · DP · FP · · · = $a^\lambda + x^\lambda$. Thus (Fig. 20), if $\lambda = 5$, $AP · CP · EP · GP · IP = a^5 - x^5$, $BP · DP · FP · HP · KP = a^5 + x^5$. Smith adds the diagram for $\lambda = 6$. This is the first published statement of this now well-known result. The first published proof is Pemberton's, and occupies pages 13–27 of *Epistola ad Amicum de Cotesii Inventis, Curvarum Ratione Quae cum Circulo & Hyperbola Comparationem Admittunt* (London, 1722).

Pemberton's proof merits Jean Bernoulli's description as 'long, tedious and intricate' (J. Bernoulli, *Opera Omnia*, vol. I (4 vols., Lausanne, 1742) no. CLX, p. 67). Pemberton made use of a result from Bernoulli on lengths of chords of multiple arcs, generalised it by incomplete induction, using a result from Brook Taylor's *Methodus Incrementorum Directa et Inversa* (London, 1716), pp. 19, 55, and derived results equivalent to the series expansion of sin $n\theta$ and cos $n\theta$.

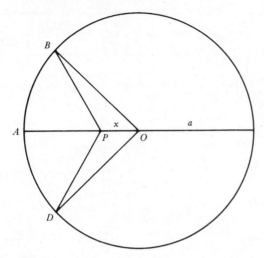

Fig. 21

Some lengthy algebra, involving extraction of the square roots of these expressions, and use of the elementary symmetric functions of the roots of polynomials, finally yielded the first published proof of the theorem, for both cases (i) P inside the circle and (ii) P outside the circle. The use of the theorem enabled Robert Smith to reorganise and extend Cotes' tables, the work which Cotes had in hand at the time of his death. The results were organised into ninety-four tables, arranged in eleven series. They include results equivalent to Cotes' original eighteen tables, and were published as Part IV of *Harmonia Mensurarum*.

Robert Smith, writing to Newton in December 1718, said: 'My Cousin Cotes, in his letter to Mr Jones, did not express himself so fully as to the content of the method as he might have done' (Cambridge University Library MS Add 3983 40). And again, on 20 August 1720: 'The tables of fluents which I calculated myself and am going to print, are to make the fourth and last part of *Harmonia Mensurarum*' (Cambridge University Library MS Add 3983 41). The tables, as published, present a formidable array of mathematical symbolism, softened a little by the excellent diagrams and quite beautiful typography (see Figs. 22 and 25).

The method of applying Cotes' theorem to the integration of rational forms is as follows:

(i)   Express the denominator in the form $x^n \pm a^n$.

(ii) With $OA = a$, $OP = x$ (Fig. 21), draw a circle centre O and radius a, and divide the circumference into $2n$ equal parts.

θ	**FORMA XXXII.**	$\dfrac{d\ddot{z}z^{6n+\frac{2}{3}n-1}}{e+fz^n+gz^{2n}}$
&c.		
3.	$\dfrac{3dz_{\frac{1}{3}n}}{5ng} - \dfrac{3dfz_{\frac{2}{3}n}}{2ngg} + \dfrac{3dp^3}{ng^3}$ ① $- \dfrac{3dq^3}{ng^3}$ ②	
2.	$\dfrac{3dz_{\frac{2}{3}n}}{2ng} - \dfrac{3dpp}{ngg}$ ① $+ \dfrac{3dqq}{ngg}$ ②	
1.	$\dfrac{3dp}{ng}$ ① $- \dfrac{3dq}{ng}$ ②	
0.	$\dfrac{-3d}{n}$ ① $+ \dfrac{3d}{n}$ ②	
$\overline{1.}$	$\dfrac{-3d}{nez_{\frac{1}{3}n}} + \dfrac{3dg}{np}$ ① $- \dfrac{3dg}{nq}$ ②	
$\overline{2.}$	$\dfrac{-3d}{4nez_{\frac{2}{3}n}} + \dfrac{3df}{neez_{\frac{1}{3}n}} - \dfrac{3dgg}{npp}$ ① $+ \dfrac{3dgg}{nqq}$ ②	
&c.		

$$a=\sqrt{ff-4eg}\,; \quad p=\tfrac{1}{2}f+\tfrac{1}{2}a\,; \quad q=\tfrac{1}{2}f-\tfrac{1}{2}a.$$

Caſ. 1.
$$OP:OA::\sqrt[3]{z^n}:\sqrt[3]{\frac{p}{g}}\,; \quad ①=\frac{1}{3a}\sqrt[3]{\frac{g}{p}} \text{ in } \beta\,(BP:BO)$$
$$-(DP:DO)+\psi(PBO).$$
$$OP:OA::\sqrt[3]{z^n}:\sqrt[3]{\frac{q}{g}}\,; \quad ②=\frac{1}{3a}\sqrt[3]{\frac{g}{q}} \text{ in } \beta\,(BP:BO)$$
$$-(DP:DO)+\psi(PBO).$$

Caſ. 2.
$$OP:OA::\sqrt[3]{z^n}:\sqrt[3]{\frac{-p}{g}}\,; \quad ①=\frac{-1}{3a}\sqrt[3]{\frac{-g}{p}} \text{ in } -(AP:AO)$$
$$+\beta(P:CO)-\psi(PCO).$$
$$OO:OA::\sqrt[3]{z^n}:\sqrt[3]{\frac{-q}{g}}\,; \quad ②=\frac{-1}{3a}\sqrt[3]{\frac{-g}{q}} \text{ in } -(AP:AO.)$$
$$+\beta(CP:CO)-\psi(PCO).$$

Fig. 22. Table for Cotes–Smith Form XXXII. This and Fig. 25 show the high quality of the work of the Cambridge University Press in 1722. From *Harmonia Mensurarum*.

(iii) Use the theorem to express $x^n \pm a^n$ as the product of such factors as PA and/or PC, i.e., $a+x$ and/or $a-x$, together with such factors as PB.

(iv) Form the product of each factor such as PB with its reflection in AO, i.e., PD, to yield a real quadratic factor $a^2 - 2ax \cos B\hat{O}P + x^2$.

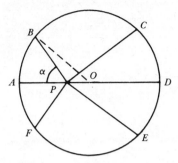

Fig. 23

(v)   The denominator now being expressed as the product of linear and/or quadratic factors, the rational form to be integrated can be expressed as the sum of partial fractions with linear or quadratic denominators, and each of these integrated as logarithms or angles or, in Cotes' terms, as the measures of ratios or the measures of angles.

As an illustration, consider the Cotes–Smith Form XXXII (Fig. 22) for $\theta = 0$.

$$\frac{d\dot{z}\,z^{\frac{2}{3}n-1}}{e+fz^n+gz^{2n}} = -\frac{d}{a}\left\{\frac{\dot{z}\,z^{\frac{2}{3}n-1}}{z^n+p/g} - \frac{\dot{z}\,z^{\frac{2}{3}n-1}}{z^n+q/g}\right\},\tag{1}$$

where $a = \sqrt{f^2-4eg}$, $p = (f+a)/2$, $q = (f-a)/2$. Put $z^n = x^3$, $p/g = b^3$ and we have

$$-\frac{d}{a}\frac{3}{\eta}\frac{x\dot{x}}{x^3+b^3}$$

for the first term. This can of course be resolved into partial fractions and integrated at once. However, to illustrate the use of the theorem in a fairly simple case, this expression becomes, using a circle of radius b (Fig. 23),

$$-\frac{d}{a}\frac{3}{\eta}\frac{x\dot{x}}{PB\cdot PD\cdot PF},\tag{2}$$

where $OP:OA::x:b$

$$= \sqrt[3]{z^n}:\sqrt[3]{\frac{p}{g}},$$

as given by Smith. Thus the first term of (1) is

$$-\frac{d}{a}\frac{3}{\eta}\frac{x\dot{x}}{PD\cdot PB^2} = -\frac{d}{a}\frac{3}{\eta}\frac{x\dot{x}}{(b+x)(b^2-bx+x^2)}$$

$$= -\frac{3d}{\eta}\frac{1}{3a}\sqrt[3]{\frac{g}{p}}\left\{\frac{x\dot{x}}{b^2-bx+x^2} + \frac{b\dot{x}}{b^2-bx+x^2} - \frac{\dot{x}}{b+x}\right\}.$$

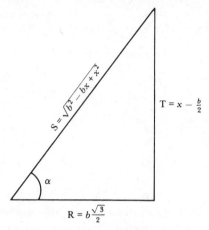

$$T = x - \frac{b}{2}$$

$$S = \sqrt{b^2 - bx + x^2}$$

$$R = b\frac{\sqrt{3}}{2}$$

Fig. 24. RTS triangle for Cotes–Smith Form XVII. From *Harmonia Mensurarum*, Cambridge University Press, 1722.

The fluents of each of the three terms in the bracket were readily available, for example, from Cotes' Forms XIII for $\theta = 0$, XIII for $\theta = 1$ and Form I for $\theta = 0$ respectively. Denoting the three terms in the bracket by (a), (b) and (c) respectively, the fluents are:

from (a) $\frac{1}{2}|(b^2 - bx + x^2) - \dfrac{b}{2\sqrt{q}}\alpha,$

from (b) $-b/q,$

from (c) $-1|(b + x),$

where α is the angle denoted by $R\left|\dfrac{R+T}{S}\right.$ (Fig. 24). Collecting these together, and evaluating between limits O and x we obtain, in the notation of Form XXXII,

$$-\frac{3d}{\eta}\frac{1}{3a}\sqrt[3]{\frac{g}{p}}\left\{\frac{1}{2}\left|\frac{b^2 - bx + x^2}{b}\right. - 1\left|\frac{b+x}{b}\right. + \sqrt{3}[-\alpha - (-P\hat{O}B)]\right\}$$

$$= -\frac{3d}{\eta}\left\{\frac{1}{3a}\sqrt[3]{\frac{g}{r}}[1|BP:PO - 1|DP:DO + \sqrt{3}\,P\hat{B}O]\right\}$$

$$= -\frac{3d}{\eta}(1),$$

although (1) is written in Smith's notation as

$$\frac{1}{3a}\sqrt[3]{\frac{g}{p}}\{\beta(BP:PO) - (DP:DO) + \psi(P\hat{B}O)\}.$$

The second term of the original fluxional form (1) yields $+3d/\eta(2)$ and, combining the two results, we have the complete integral.

Notes

(i) Care is needed with the sign of $x - \frac{1}{2}b$, but it will be seen that the final result is always angle *PBO*.

(ii) The moduli of the measures (β, ψ in this case), are always the lengths of chords drawn from points such as *B*, parallel to *OA* for the ratios, and perpendicular to *OA* for the angles, the radius always being regarded as unity.

Thus, in the table for Cotes–Smith Form XVII (Fig. 25), $\beta, \gamma, \delta, \ldots$ are lengths of chords from points B, C, D parallel to AO, and $\psi, \chi, \phi, \ldots$ are lengths of chords from the same points perpendicular to AO, the radius being regarded as unity.

Among the most vocal of those urging Robert Smith to get on and publish Cotes' work was Brook Taylor, Secretary of the Royal Society (see Brook Taylor to Professor Smith, 27 November 1718 and 11 December 1718 [4], Letters CXIX and CXX). In 1718, Taylor used one of Cotes' fluxional forms as a challenge to the Continental mathematicians. The challenge, submitted through Pierre Remond de Montmort (d. 1719 – see *Biographie Universelle*) was to find the fluent of

$$\frac{\dot{z}z^{(\delta/\lambda)q-1}}{e + fz^q + gz^{2q}},$$

where δ was a positive or negative integer, and λ a power of 2. This is the Cotes–Smith Form XXXII for $\theta = 0$. The exact terms of the challenge can be read in Jean Bernoulli's reply in the *Acta Eruditorum* for June 1719, p. 256: 'To show how the fluent of $\dot{x}/(a^4 + x^4)$ could be related to the quadrature of the circle, contrary to what Leibniz had suggested, and further, to extend the solution of the first problem to fluxions of the type

$$\dot{z}z^{(\delta/\lambda)q-1} : (e + fz + gz^{2q} + hz^{3q}),$$

where λ is any one of the series 2, 4, 8, 16, ...' Bernoulli commented tartly that Leibniz had not denied the possibility of reducing $\int \dot{x} : (a^4 + x^4)$ (Bernoulli's and Leibniz' notation) to the quadrature of the circle and of the hyperbola (i.e., evaluating the integral in terms of logarithms and angles), but only denied the possibility of so doing by the analysis he had used. He had, said Bernoulli, overlooked the

θ	F O R M A XVII. $\dfrac{d\ddot{z}z^{en-\frac{1}{7}n-1}}{e+fz^n}$
&c.	
4.	$\dfrac{7dz^{\frac{2}{7}\pm n}}{24nf} - \dfrac{7dez^{\frac{17}{7}n}}{17nff} + \dfrac{7deez^{\frac{10}{7}n}}{10nf^3} - \dfrac{7de^3z^{\frac{3}{7}n}}{3nf^4} + \dfrac{7de^4}{nf^5}$ Ⓘ
3.	$\dfrac{7dz^{\frac{17}{7}n}}{17nf} - \dfrac{7dez^{\frac{10}{7}n}}{10nff} + \dfrac{7deez^{\frac{3}{7}n}}{3nf^3} - \dfrac{7de^3}{nf^4}$ Ⓘ
2.	$\dfrac{7dz^{\frac{10}{7}n}}{10nf} - \dfrac{7dez^{\frac{3}{7}n}}{3nff} + \dfrac{7dee}{nf^3}$ Ⓘ
1.	$\dfrac{7dz^{\frac{3}{7}n}}{3nf} - \dfrac{7de}{nff}$ Ⓘ
0.	$\dfrac{7d}{nf}$ Ⓘ
$\overline{1}.$	$\dfrac{-7d}{4nez^{\frac{4}{7}n}} - \dfrac{7d}{ne}$ Ⓘ
$\overline{2}.$	$\dfrac{-7d}{11nez^{\frac{11}{7}n}} + \dfrac{7df}{4neez^{\frac{4}{7}n}} + \dfrac{7df}{nee}$ Ⓘ
$\overline{3}.$	$\dfrac{-7d}{18nez^{\frac{18}{7}n}} + \dfrac{7df}{11neez^{\frac{11}{7}n}} - \dfrac{7dff}{4ne^3z^{\frac{4}{7}n}} - \dfrac{7dff}{ne^3}$ Ⓘ
$\overline{4}.$ *&c.*	$\dfrac{-7d}{25nez^{\frac{25}{7}n}} + \dfrac{7df}{18neez^{\frac{18}{7}n}} - \dfrac{7dff}{11ne^3z^{\frac{11}{7}n}} + \dfrac{7df^3}{4ne^4z^{\frac{4}{7}n}} + \dfrac{7df^3}{ne^4}$ Ⓘ

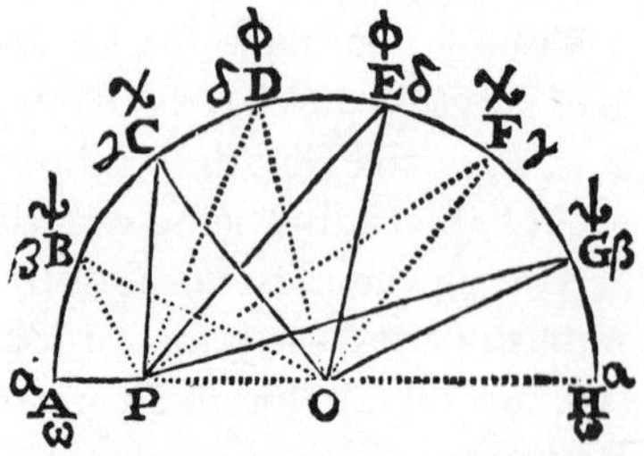

Caſ. 1. $OP:OA::\sqrt[7]{z^n}:\sqrt[7]{\dfrac{e}{f}}$;

Ⓘ $= \frac{1}{7}\sqrt[7]{\dfrac{f^4}{e^4}}$ in $-\delta(BP:BO) +\gamma(DP:DO) -\beta(FP:FO)$ $+(HP:HO) +\phi(PBO) -\chi(PDO) +\psi(PFO)$

Caſ. 2. $OP:OA::\sqrt[7]{z^n}:\sqrt[7]{\dfrac{-e}{f}}$;

Ⓘ $=-\frac{1}{7}\sqrt[7]{\dfrac{f^4}{e^4}}$ in $-(AP:AO) +\beta(CP:CO) -\gamma(EP:EO)$ $+\delta(GP:GO) +\psi(PCO) -\chi(PEO) +\phi(PGO)$

Fig. 25. Table for Cotes–Smith Form XVII. From *Harmonia Mensurarum*, Cambridge University Press, 1722.

possibility of expressing $dx : (a^4 + x^4)$ as

$$(1 : \sqrt{2}a + x; 2a\sqrt[3]{2})\, dx : (a^2 + ax\sqrt{2} + x^2)$$
$$+$$
$$(1 : \sqrt{2}a - x; 2a\sqrt[3]{2})\, dx : (a^2 - ax\sqrt{2} + x^2),$$

i.e., in modern notation

$$\frac{1}{2a\sqrt[3]{2}} \left\{ \frac{\sqrt{2}a + x}{a^2 + ax\sqrt{2} + x^2} + \frac{\sqrt{2}a - x}{a^2 - ax\sqrt{2} + x^2} \right\} dx,$$

and thus effecting the integration by two quadratures of the circle and two of the hyperbola.

Bernoulli then turned to the main problem of integrating

$$\frac{\dot{z}z^{(\delta/\lambda)q - 1}}{e + fz^q + gz^{2q}}.$$

He had clearly nothing like Cotes' general method available, but in twenty concise and carefully argued propositions, came near to achieving it. By substituting $z^q = y = x^\lambda$, Bernoulli reduced the integrand to

$$\frac{x^{\delta - 1}}{e + fx^\lambda + gx^{2\lambda}}\, dx.$$

There are two cases, according as f^2 is greater or less than 4 *eg* and, in both cases, the denominator can be replaced with two trinomials of the form $(a + bx^{\frac{1}{2}\lambda} + cx^\lambda)(a - bx^{\frac{1}{2}\lambda} + cx^\lambda)$, and the process repeated until only linear or irreducible quadratic factors remain. Clearly, λ must be a power of 2 for the method to succeed. The method is equivalent to factorisation by Cotes' theorem, which suffers from the same restriction. Bernoulli neatly extended the argument to complete the integration of

$$z^{(\delta/\lambda)q - 1}\, dz : (e + fz^q + gz^{2q} + hz^{3q}),$$

reducing the denominator to $(A + Bx^\lambda)(C + Dx^\lambda + Ex^{2\lambda})$, each of which factor can be further factorised as before.

Jacob Hermann, Jacques Bernoulli's pupil, also published a reply to Brook Taylor's challenge (*Acta Eruditorum* (August 1719), p. 351) and proved the general truth (which Bernoulli had not done) of the method of reducing a trinomial to two further trinomials.

We see then that, before Cotes' work was published in 1722, the Continental mathematicians were able to develop methods of comparable power, with, in general, a more suitable notation. It was indeed appropriate that the work should have been rounded out and completed by the Anglo–Frenchman De Moivre. In *Miscellanea Analytica*

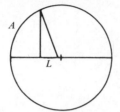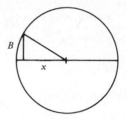

Fig. 26. Arc A is n times arc B.

(London, 1730), De Moivre began with a statement, without proof, of the theorem which bears his name, in the form

$$x = \frac{1}{2}\sqrt[n]{L + \sqrt{(L^2 - 1)}} + \frac{1}{2}\frac{1}{\sqrt[n]{L + \sqrt{(L^2 - 1)}}},$$

where x is the cosine of an arc B, and L is the cosine of n times that arc, i.e., A (Fig. 26). This is equivalent to

$$\cos\theta = \tfrac{1}{2}[(\cos n\theta + \mathrm{i}\sin n\theta)^{1/n} + (\cos n\theta - \mathrm{i}\sin n\theta)^{1/n}].$$

From this De Moivre deduced, in ten short corollaries, the factorisation of $1 \pm Z^n$ for n integral and positive, odd or even, obtaining first the general resolution of $Z^{2n} - 2LZ^n + 1$ into quadratic factors, and then the special cases for $L = \pm 1$. The factorisations given by De Moivre were, in modern terms,

$$1 - Z^n = (1 - Z^2)\prod_{r=1}^{\frac{1}{2}n - 1}\left(Z^2 - 2Z\cos\frac{2r\pi}{n} + 1\right), \quad n \text{ even},$$

$$1 - Z^n = (1 - Z)\prod_{r=1}^{\frac{1}{2}(n-1)}\left(Z^2 - 2Z\cos\frac{2r\pi}{n} + 1\right), \quad n \text{ odd},$$

and the corresponding results for $1 + Z^n$. So the first published application of De Moivre's theorem was to prove Cotes' theorem, in the process giving the general factorisation of $x^{2n} - 2Lx^n + 1$, and hence removing the necessity for λ to be a power of 2 in the type of problem sent to Montmort.

De Moivre went on to develop applications of the theorem to expressing rational functions in partial fractions, thereby further widening the range of such forms which could be integrated. Thus, De Moivre's work in *Miscellanea Analytica* was, as previously noted, a rounding and completing of Cotes' work on integrals, and a preparation for the more analytical approach of Euler and others throughout the succeeding years of the eighteenth century. The factorisations given above remain in the literature as the Cotes–De Moivre properties of the circle.

Other proofs of Cotes' theorem appeared from time to time. Two merit brief mention. Thomas Simpson gave a proof in *The Doctrine and Applications of Fluxions*, vol. II (2 vols., London, 1750), p. 347, and James Brinkley, in the *Transactions of the Royal Irish Academy*, 7 (1797), 151–9, gave a proof depending 'on the circle only', i.e., like Pemberton before him, but for different reasons, avoiding the use of complex numbers. Brinkley claimed that previous proofs, from the circle only, had assumed the expansion of $\cos n\theta$ as a power series in $\cos \theta$ (the result which Pemberton had from Taylor, by incomplete induction), but this had not been proved until 1760 by E. Waring (Lucasian Professor 1760–98) in *Letter to the Reverend Dr Powell* (Cambridge, 1760). If this claim is granted, and it seems that it should be, we must acknowledge Brinkley's as the first complete proof, 'from the circle only', unless, as seems unlikely, Cotes had one. Simpson's 1750 proof was somewhat less general.

5

Astronomy

On 13 February 1709/10 Richard Bentley wrote of 'The College Gatehouse rais'd up and improv'd to a stately Astronomical Observatory, well stor'd with the best instruments in Europe' (R. Willis & J. W. Clarke, *The Architectural History of the University of Cambridge and Colleges of Cambridge and Eton* (Oxford, 1886), p. 500). This is Bentley's typical exuberance; the observatory was not finally completed until 1739, although of course it was in use from about the time of Bentley's remark. The appearance of the King's gate of Trinity in 1740 is shown in Fig. 27.

Cotes, assuming his Professorial duties in his mid-twenties, set about them with energy and enthusiasm. Money, for constructing the observatory and purchasing instruments, was raised by subscription. Organising the subscription list, paying the workmen and generally supervising the work were among the duties which Bentley assigned to Cotes. Yet by 25 April 1715, the observatory was without a satisfactory clock (see the letter from Cotes to Newton, 29 April 1715, below, part of which is in [4], Letter xc).

It is probable that practical astronomy, as distinct from tuition at undergraduate level, did not figure very largely in the programme of work at the observatory. The only records of observations which have so far come to light are Cotes' report of his observations of the total eclipse of the sun on 22 April 1715, recorded in his letters to Newton of 29 April and 13 May of that year and the letter of 15 March 1716 to Dr Dannye, rector of Spofforth in Yorkshire, describing what is generally thought to have been an auroral display observed on 6 March 1715/16. This letter is preserved in the original letter book of the Royal Society in London, and was published in *Philosophical Transactions of the Royal Society* 31 (1720), 60.

Cotes' letter to Newton, of 29 April 1715, follows. This letter included Cotes' undimensioned sketch of his device for taking altitudes

80

Fig. 27. The Great Court at Trinity College, Cambridge, in 1740, showing the observatory in place over the King's gate. From R. Willis & J. W. Clarke, *The Architectural History of the University of Cambridge and the Colleges of Cambridge and Eton*, Oxford, 1886.

(Fig. 28) in connection with determining solar mid-day by the method of equal altitudes. It is a simple device for bringing the telescope quickly back to a given elevation, and presumably freeing it in between whiles for other uses. The need to construct such a piece of apparatus as Cotes describes which on the face of it looks a little crude, suggests that the observatory was not particularly well equipped.

Cambridge April 29th 1715.

Sir

I think it my duty to send You what Observations I could make of the last Eclipse.

Times by y^e clock	Times corrected	
$8^h 07' 35''$	$8^h 10' 09''$	About 30 degrees in the Sun's Limb covered by the Moon, estimated by memory after the end of the Eclipse.
8 31 38	8 34 11	The first edge of the greater Spot touched by the Moon.
8 32 42	8 35 15	The first edge of the middle Spot touched.
8 34 22	8 36 55	The first edge of the third Spot touched.
9 12 04	9 14 37	The first recovery of the Sun's Light.
10 19 25	10 21 57	The end of the Eclipse.

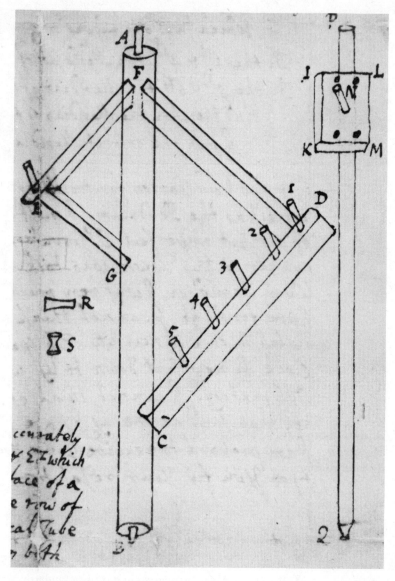

Fig. 28. Cotes' sketch of his device for applying the method of equal altitudes. Cambridge University Library MS Add 3983 39. (By permission of Cambridge University Library.)

The Times were corrected by an instrument which I ordered to be made for the purpose about a day or two before the Eclipse: the form of which is as follows. *AB* is a strong wooden Axis of about six feet in length, *CD* and *DF* on one side, *FE* and *EG* on the other are pieces fram'd to each other & to ye axis as firmly as was possible. Into the piece *CD* & at the angle *E* were fix'd strong wooden pins nearly parallel to each other & perpendicular to ye plain *CDFEG*. *PQ* is the Cylindrical Brass-Tube of a Five-foot Telescope, this was well fastened with Iron staples & screws to ye piece of wood *IKML* whose under plain surface is here represented as objected to view. Into this surface there were perpendicularly fix'd a strong wooden Pin *N* which was design'd to hang the upper end of the Telescope upon the Pin *E* whilst the lower end rested upon any of the Pins *CD*. Now that ye Telescope might be taken off, and yet afterwards be again plac'd accurately in the same position, I ordered the edges *JK* and *EF* which touch'd each other to be rounded like the surface of a Cylinder, as also the edge *CD* into which the row of Pins was fix'd and against which the Cylindrical Tube of the Telescope rested, so that the contact in both places might be made in a point. Upon the same account, the Pin *E* was made a little hollow as is represented at *R*,; the others were Frustums of Cones, that thereby the Telescope might more surely touch the edges *EF* and *CD*. Into the two ends of the wooden Axis were strongly fix'd two pieces of well temper'd steel: that at the upper end *A* was a Cylinder well turn'd which moved in a Collar whose circumference (represented by *S*) was figured like two equal hollow & inverted Frustums of Cones joined together: the lower at *B* was a cone moving in a Conical Socket of a somewhat bigger angle. This Socket had liberty to move horizontally & to be fix'd in any Position by three Screws which press'd against it sideways. The Instrument being thus prepar'd, I fixed a Needle at the lower end of the Wooden Axis whose point stood out from it about an Inch. Then suspending a fine Plumb-line from the uppermost end of the same Axis I altered the Position of the Instrument by the three Screws, until the Plumb-line came to rest against the point of the Needle in the whole revolution of the Instrument, & there I fix'd it as prepared for use. My Observations follow.

	Day		
1.	XXI pm	$4^h 01' 21''$	the Sun's upper Limb observ'd at the third Pin.
2.	XXII am	6 48 41	Upper Limb ⎫ 2^d Pin
		6 52 09	Lower Limb ⎭
3.	XXIII am	6 47 29	Upper Limb ⎫ 2^d Pin
		6 50 58	Lower Limb ⎭
4.	XXIII am	7 51 10	Upper Limb 3^d Pin
5.	XXV am	6 44 53	Upper Limb ⎫ 2^d Pin
		6 48 22	Lower Limb ⎭
6.	XXV pm	5 08 18	Lower Limb ⎫ 2^d Pin
		5 11 47	Upper Limb ⎭

Allowing for the Variation of Declination I find by the 2^d and 3^d observ: the length of Solar Day measured by the Clock was $24^h 00' 18''$

By the 3^d and 5th observ: the length of two Solar Days by the Clock was 48^h 00′ 18″

Which two Conclusions manifest a great inequality of the Clock's motion

By the 1st & 4th the Meridian of the XXII^d day was at   11^h 57′ 32″
By the 5th & 6th the Meridian of the XXVth day was at   11^h 58′ 02″
And therefore the Meridian of the XXII^d, allowing for the
Clock's inequality   11^h 57′ 26″
I put the correct Meridian of the XXII^d day at   11^h 57′ 29″

I beg Your Pardon for troubling You with so large an account of my method for correcting the Pendulum. I must confess to You I have a design in it for the advantage of our yet imperfect Observatory. The Clock which I used was borrowed of a Clock-maker in this Town, who took it for a very good one. Not expecting so great inequality in its motion, I was very much surprised to find it by the Observations: & since I have found it I cannot think of making use of such ordinary workmanship again unless in case of necessity. To speak plainly, I beg of You to let that excellent clock be now sent down to us which You ordered to be made for the use of the Observatory. I cannot think of a more accurate Instrument for the setting of it than such an one as I have been describing. Having it therefore by me, I think I am prepared to receive Your Noble Gift. I have written to Mr Street to wait upon You for Your resolution.

I am Sir
Your
Most Obliged Humble Servant
Roger Cotes.

I will send you an account of what was observed at Cambridge during the total obscuration in another Letter.

The letter of 29 April is reproduced in part in [4], from the incomplete draft in Trinity College, which lacks the description of the instrument. Newton made a positive response to Cotes' appeal for the promised clock and Cotes was able to begin his letter to Newton of 13 May 1715 containing the promised observations of the eclipse with the words: 'Mr Bentley has told me, You have been pleas'd to give orders, that the Clock may be sent to Cambridge. I take this opportunity of returning you my hearty thanks for it' [Fig. 29] (Cotes to Newton, 13 May 1715 [4], Letter XCI).

Of the two drawings shown (Fig. 30(a) and (b)), the first is Cotes' own drawing as it appeared to him at the time of the sun's total obscuration, and shows a ring of white light surrounding the moon's body, and rays of paler light in the form of a rectangular cross, the longer arm of the cross lying very nearly along the ecliptic. The second drawing, says Cotes, is by a very ingenious gentleman, representing the appearance as seen by himself. Apart from their intrinsic interest

Fig. 29. The clock (now in the Master's lodge at Trinity College, Cambridge) presented by Newton to the Trinity observatory. (By permission of the Master.)

as showing the sun's corona, they merit inclusion here as being among the very few records of astronomical observations made by Cotes or, indeed, by any one else at the Trinity observatory during Cotes' Professorship.

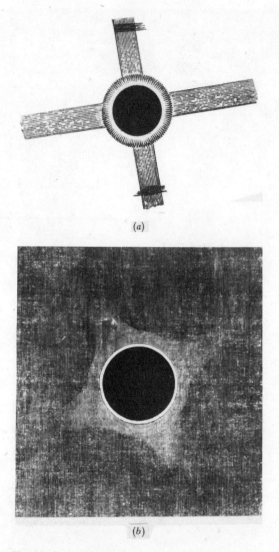

Fig. 30. Cotes' sketch of the total eclipse of the sun, 22 April 1715. From J. Edleston, *Correspondence of Sir Isaac Newton and Professor Cotes*, London, 1850. (By permission of Frank Cass and Co. Ltd.)

The accurate determination of time by the method of corresponding altitudes, which Cotes used, seems to have been a fairly recently developed technique. J. Delambre attributes it to J. Picard, at the Paris Observatory, on 2 February 1674, commenting that Picard knew nothing of the differential formula (see J. B. J. Delambre, *Astronomie Théorique et Pratique*, vol. I (3 vols., Paris, 1814) p. 571). The method requires an accurate clock or, as in the results described by Cotes, can be used to check the accuracy of a clock. Delambre gives a very full account of the method in his own work *Astronomie Théorique et Pratique*, Chapter XIX, together with differential formulae connecting the change in hour-angle with the change in declination; he also gave solar tables calculated for the latitude of Paris. Cotes probably used similar methods, most likely his Theorem XXIII from his tract, Aestimatio Errorum, to be discussed below, and the source, via La Caille, of Delambre's formulae. Cotes took a very serious interest in the practical aspects of astronomy and was concerned with instrumentation, with improving the accuracy of observations, and with developing suitable techniques of calculation. In a letter to his uncle John Smith, of 10 February 1707/8 ([4], Letter XCVIII), Cotes said he had been in London to view the large brass sextant which had been made for the College by Rowley at a cost of one hundred and fifty pounds, that Sir Isaac had given orders for the making of a pendulum clock as a gift to the College, at a probable cost to him of fifty pounds, and that he himself had 'lately hit upon a device which I beleive will be of very good use for observing eclipses'. The sextant is illustrated and described in J. Harris, *Lexicon Technicum*, vol. II (2 vols., London, 1723), Introduction. It was a splendid instrument and is shown in Fig. 31. John Harris says he included the illustration in *Lexicon Technicum*: 'to do further justice to our exalted mathematical Instrument Maker Mr John Rowley, in Johnson's Court in Fleet Street', and that the instrument, 'for its universal use, far exceeds any mathematical instrument ever yet made'.

The clock given by Newton, and which arrived some seven years later, is preserved in the Master's Lodge at Trinity and is illustrated in Fig. 29. Cotes' device, which he said he had 'lately hit upon', was an equatorially mounted telescope, fixed parallel to the earth's axis of rotation, and rotating at the same speed as the earth. By suitably adjusting a mirror or prism fixed to the telescope, a particular object could be kept in view for any required length of time (see Fig. 32). There had been several anticipations of this idea towards the end of the seventeenth century; some are listed in, for example, A. Wolf, *A History of Science Technology and Philosophy in the XVIIIth Century* (London, 1938). Cotes, if not original in his idea, was certainly very up to date, and developed the idea more fully in the following unpublished 'Clare paper':

A paper of Mr Professor Cotes's, communicated by M. Folkes Esq F.R.S.

Tis proposed that for taking the time of any transit over the Meridian a Telescope be provided wch shall revolve about its axis within 2 fix'd collars *ab*

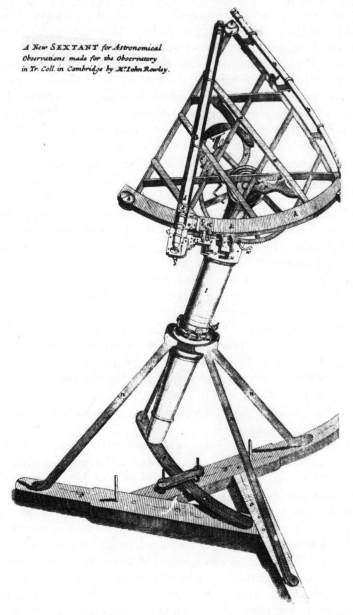

A New SEXTANT for Astronomical Observations made for the Observatory in Tr. Coll. in Cambridge by Mr John Rowley.

Fig. 31. The Trinity sextant. From J. Harris, *Lexicon Technicum*, London, 1723.

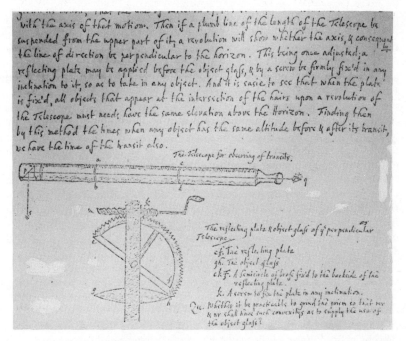

with the axis of that motion. Then if a plumb line of the length of the Telescope be suspended from the upper part of it, a revolution will show whether the axis, & consequently the line of direction be perpendicular to the horizon. This being once adjusted, a reflecting plate may be applied before the object glass, & by a screw be firmly fix'd in any inclination to it, so as to take in any object. And it is easie to see that when the plate is fix'd, all objects that appear at the intersection of the hairs upon a revolution of the Telescope must needs have the same elevation above the Horizon. Finding then by this method the times when any object has the same altitude before & after its transit, we have the time of the transit also.

The Telescope for observing of transits.

The reflecting plate & object-glass of y perpendicular Telescope.
ef. The reflecting plate
gh. The object glass
ckF. A semicircle of brass fix'd to the backside of the reflecting plate
k. A screw to fix the plate in any inclination.
Qu. Whether it be practicable to grind the prism so that my & nv shall have such convexities as to supply the use of the object glass?

Fig. 32. Cotes' diagram for a transit telescope. (By permission of Clare College Library, Cambridge.)

& cd; That by endless screws apply'd at the collar next to the eye glass this motion be regulated so as to become as even and slow as shall be desired; that by so ordering the cross hairs the line of direction of the Telescope be made to coincide with the axis of that motion; That a triangular glass Prism, having the angle v of 90°, & the angles m & n of 45° each, be so fixt before the object glass that of the 2 sides w[ch] include the right angle the one nv be parallel to the axis of the Telescope, and the other mv be at right angles to the same. Having adjusted the prism that the reflected ray pq which coincides with the axis of the Telescope shall always be at right angles with its incident ray ps, (w[ch] may be done as accurately as the Telescope can distinguish) 'tis evident that all objects w[ch] appear at the intersection of the hairs upon a revolution of the Telescope must be situated in the same great circle. Now the best way to place the Telescope so that the great circle may be the same with the Meridian seems to be that of directing it to several objects at the times of their transits: And to find the times of those transits accurately it is further proposed;

That another Telescope of a considerable length be made to revolve about its axis, in a perpendicular posture, within two collars as before, with endless screws to govern its motion; That the direction of line of this Telescope be also made to coincide with the axis of that motion. Then if a plumb line of the length of the Telescope be suspended from the upper part of it, a revolution will show whether the axis, and consequently the line of direction be perpendicular to the horizon. This being once adjusted; a reflecting plate may be applied before the object glass, & by a screw firmly fix'd in any inclination to it, so as to take in any object. And it is easie to see that when the plate is fix'd, all objects that appear at the intersection of the hairs upon a revolution of the Telescope must needs have the same elevation above the Horizon. Finding then by this method the times when any object has the same altitude before and after its transit, we have the time of the transit also.

Of all Altitudes those are most proper for determining of time wch are taken in the Eastern and Western Verticals, if the uncertainty of Refraction hinder not, for in a given moment of time, the variation of altitude of any object is to the arch of the equator wch passes over the Meridian in the same moment; as the rectangle of the complement of ye Poles elevation & the Sine of the Object's Azimuth, to ye square of the radius. So that in a given Latitude the variation of Altitude is the same for all objects, whatever be their Declination, provided thay have the same Azimuth. But if the Azimuth be different under the same elevation of the Pole, the variation of Altitude will be as the Sine of the Azimuth.

A Telescope like the former may be placed to take the time of a star's coming to the Eastern of Western vertical; by w^{ch} & the time of its transit over the Meridian the Declination may be determined without any regard to refractions.

A Telescope like the latter may be directed to the Pole of the world; and thereby eclipses & any other such like appearances be observed more easily than is usual.

Cambridge

In the Latitude of 52° 15′ one degree of error in Altitude causes an error of 6′ 32″ in time; Although the Observation be made in Eastern or Western vertical.

To determine the time to the accuracy of 1″, the Altitude must be taken to the accuracy of 9″.

(Morgan Manuscripts, Clare College Library, Cambridge.)

Before leaving the subject of instruments it is worth remarking that Robert Smith included in his *Compleat System of Opticks in Four Books*, vol. I (2 vols., Cambridge, 1738), p. 377 and Fig. 588, a description and a more finished drawing of Cotes' corresponding altitudes instrument, remarking that 'The Earl of Islay has one such.' The cylindrical brass tube was, says Smith, 'that from our own sextant'.

For an informative and interesting article on the early instruments, some of which formed part of Cotes' equipment, see D. J. Price, 'The Early Observatory Instruments of Trinity College, Cambridge', *Annals of Science*, 8 (28 March 1952) 1. A list kept in Trinity Library is shown in Appendix 5.

Cotes' interest in practical astronomy led him, from instruments, to a concern for the accuracy of observations, and to a serious attempt to assess these errors and to minimise them. His ideas are set out in the short Latin tract, Aestimatio Errorum in Mixta Mathesi. (The term 'Mixed Mathematics', denoting mathematics used in practical applications, was in use until well into the nineteenth century.) The tract is one of several forming Cotes' Opera Miscellanea, usually bound up in one volume with *Harmonia Mensurarum*, ed. R. Smith (Cambridge, 1722). Astronomical observations commonly require the solution of plane or spherical triangles and, in his tract, Cotes presented twenty-eight theorems, in which were derived differential formulae relating the small variations in two elements of a triangle, two other elements being held constant. Theorems i–x are concerned with plane triangles, the remaining eighteen with spherical triangles. The work epitomises the elegance and conciseness of Cotes' methods, and falls somewhere between 'geometrial fluxions' and modern differential methods. Of all Cotes' published papers, it is the one which seems to have aroused most interest, particularly in France. Lalande said, correctly as far as I have been able to ascertain, that Cotes was the first to give these theorems. Lalande considered the formulae to be accurate enough for most of the purposes of practical astronomy, indeed accurate enough to cope with observational errors of up to one degree; and that if, as was common with astronomical observations, the error was a matter of minutes, the error arising from the use of the formulae could be very small indeed – Lalande mentioned 'hundredths of a second' (see J. J. Lalande, *Astronomie*, vol. iii (3 vols., Paris, 1792), p. 589). Lalande stated, in full, the theorems on spherical triangles, more or less in Cotes' words (allowing for translation from Cotes' Latin into Lalande's French), developed several alternative expressions and indicated several problems in astronomy to which the formulae could successfully be applied (see later).

Cotes' work seems to have received rather more notice in France than in England. In connection with the paper at present under discussion, two other commentators besides Lalande must be mentioned: they are the Abbé La Caille, and J. B. J. Delambre.

La Caille's Mémoire, of 1741, first gave some publicity to Cotes' Aestimatio. La Caille restated the eighteen theorems on spherical triangles (rather unnecessarily) as twenty-four, and gave with each result the value of the variables for which he considered the results to be most accurate (see Fig. 41). He spoke with approval of the great utility of the theorems in simplifying calculations which were otherwise long and tedious, and supplied some illustrative examples worked out in detail. He observed:

Having sought for some time for a simple and general method, I fell upon an excellent book of Mr Cotes entitled *Harmonia Mensurarum*, in which is found a treatise Aestimatio Errorum in Mixta Mathesi. The aim of the author is to determine the limits of unavoidable errors in geometry and astronomy, regarding them as infinitesimal differences ... After having read this treatise with pleasure I found it easy to apply the rules to such problems as parallax, refraction, precession of the equinoxes, aberrations, and all the little movements, as well as errors of observation. I have deduced from these, methods of calculation so simple, that I felt I must report some of them.

(See N. L. de La Caille, 'Calcul des Différences dans la Trigonométrie Sphérique', *Histoire de L'Académie Royale des Sciences* (Paris, 1741) p. 238.)

Delambre comments that 'since La Caille's "éloge", on Cotes' treatise, Cotes' theorems appear in all our treatises of astronomy, geometry and navigation'. With a touch of asperity, Delambre warns against abuse of the methods. The theorems must be used within the restrictions suggested by the author (i.e., La Caille–Cotes did not comment on restrictions, except to observe that the variations must be small). Assuming, for example, that the parallax of the moon, which can exceed one degree, can be treated as a differential, can lead to substantial errors in other quantities calculated from it. One must at least know what is being neglected, and our standards of accuracy are now, said Delambre, better than forty years ago.

In an earlier work, Delambre illustrated the nature of the approximations made, by restating Cotes' Theorems I–X i.e., the theorems on plane triangles, in full, and pointing out the terms which had been neglected (see Fig. 39). The interested reader could refer to the following works: J. B. J. Delambre, *Histoire de l'Astronomie au Dix-huitième Siècle* (Paris, 1827), pp. 449–57 (Delambre's *Histoire* for the eighteenth century appeared posthumously in 1827, but the author was at work on it for many years – see *Encyclopaedia Britannica*, 9th edition, entry under 'Delambre, J. B. J.'); J. B. J. Delambre, *Astronomie Théorique et Pratique*, vol. I, pp. 239–59.

Among the works cited by Delambre as using Cotes' theorems, are, for example, Moreau de Maupertuis, *Astronomie Nautique* (Paris, 1743). This excellent and concise little work, written with admirable French 'clarté', contains the differential formulae, exemplified for use by an observer in a fixed observatory, and also by an observer at sea. No reference is made to Cotes, even in the long Preface. Among other authors cited by Delambre are Maudit, *Trigonométrie*; Boscovitch, *Ceuvres* (vol. IV); Cognoli, *Trigonométrie*.

Cotes, in his own Introduction to the treatise, said that astronomy depended on accurate and careful observation of the heavens. The

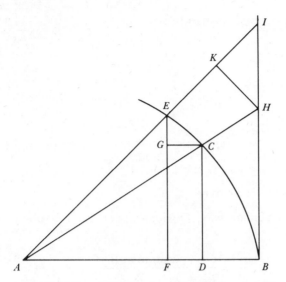

Fig. 33. *CE* and *EG* are the small variations considered. From *Harmonia Mensurarum*, Cambridge University Press, 1722.

limitations of our senses and the imperfections of our instruments led to errors which we could never wholly avoid. We could however set limits to the errors, and it was important to have a system whereby we could define these limits, and know what errors were negligible. If we were aware of these, we would proceed surely and by design, not led by blind fortune.

Cotes begins Aestimatio Errorum with three lemmas. I give the first in detail to illustrate the style, the others briefly. The small variations of lines or angles are referred to throughout by Cotes as 'minimum variationes' – and he is always careful to give them in the form of a ratio one to the other. They are probably 'geometrical fluxions' to Cotes, but are equally well and more conveniently regarded as differentials.

Lemma I (Fig. 33)
The small variation of any arc of a circle is to the small variation of the sine of that arc, as the radius to the sine of the complement.

Proof
'With centre *A*, radius *AB*, arc *BC* is drawn; *CG* is parallel to the radius *AB*; *CE* is the variation of arc *BC*; *EG* is the variation in the sine of that arc. I say *CE*:*EG*::*AC*:*AD*, when the variations are very small.' In this case, triangles *CEG*, *CAD* are similar, therefore

$CE : EG :: AC : AD$, i.e., Lemma I proves

$$\frac{d(r\theta)}{d(\sin \theta)} = \frac{r}{\cos \theta}.$$

In Lemmas II and III the results

$$\frac{d(r\theta)}{d(r \tan \theta)} = \frac{1}{\sec^2 \theta}$$

and

$$\frac{d(r\theta)}{d(r \sec \theta)} = \frac{1}{\sec \theta \tan \theta}$$

are proved in a similar way. Obviously, Lemma I is obtained by assuming

$$\frac{\sin (\text{variation in } B\hat{A}C)}{2} = \frac{\text{variation in } B\hat{A}C}{2},$$

i.e., the first term (ignored in the approximation is of the order of (variation in $B\hat{A}C)^3$ and, throughout the treatise, results are obtained by similar approximations, involving sometimes the squares of the variations. In Lemma I the accuracy of the approximation does not depend upon the value of θ, but this is not always the case, as will be shown.

Cotes now develops ten theorems on plane triangles, in which two of the six elements remain constant, and the ratio of small variations in two others are calculated. They are all obtained by synthetic methods with Cotes' usual elegance and economy and he adds a note that since angles and lines are different kinds of quantities, their variations cannot have a ratio to each other, therefore, we must use a linear measure for angles, i.e., a circular arc. The angular measure is therefore in 'radians' throughout. I have set out the results in tabular form, Fig. 39, in which the symbol r refers to the radius of the canon used, which, without loss of generality, can be taken as unity. It should also be noted that Cotes does not, in general, use cosines, cotangents and cosecants, preferring to speak of sines, tangents and secants of complementary arcs.

As illustrations of the methods, I give below the details of Cotes proofs of Theorem II and of Theorem VIII:

Theorem II (Fig. 34)
One angle and a side adjacent to it remaining constant, the measure of the small variation of either of the other angles is to the small variation of the

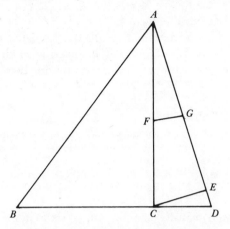

Fig. 34. Small variations in the angle A and the side AC of triangle ABC are considered. From *Harmonia Mensurarum*, Cambridge University Press, 1722.

side opposite the fixed angle, as the tangent of the angle opposite the fixed side is to the side opposite the fixed angle. With centre A and radius AF, that of the given canon, let FG, the measure of the variation of angle A, which is also the measure of the variation of angle C, be described. I say, this measure FG to be to DE, the variation of the side AC which is opposite the constant angle B, as the tangent of the angle C, which is opposite the constant side AB, is to the side AC, which is opposite the constant angle B, where the variations are very small. Now in this case FG will be to CE as the radius to AC, and CE to DE as the tangent of the angle CDE or C to the radius: therefore, mix the equalities, FG will be to DE as the tangent of the angle C to AC.

Summarising this: AB, B, being constant,

$$\frac{d(A)}{d(AC)} = \frac{\tan C}{AC},$$

where the variations are small.

Proof (summarised)

$$FG:CE::r:AC \quad (r = AF = \text{radius of canon}). \tag{1}$$

$$CE:DE::\tan CDE:r::C:r. \tag{2}$$

From (1) and (2)

$$FG:DE::\tan C:AC, \quad \text{i.e.,} \quad \frac{d(A)}{d(AC)} = \frac{\tan C}{AC}.$$

The construction, described in Theorem I, requires AE to be equal to AC, and Cotes obtains his approximate results by also using CE as the perpendicular from C to AD. The error involved is small if, as Cotes reiterates, the variations are small. From the sine rule in triangle CED, where $CE = 2AC \sin \frac{1}{2}A$,

$$D = C - \delta A, \quad E\hat{C}D = 90° + \frac{1}{2}\delta A - C,$$

we obtain the correct result, agreeing with that given by Delambre (Fig. 39), i.e.,

$$2 \sin \tfrac{1}{2}\delta A = \frac{\delta(AC) \sin (C - \delta A)}{AC \cos (C - \frac{1}{2}\delta A)}.$$

The (computer calculated) results (see Fig. 40) give the percentage variation in AC consequent upon a given variation in the angle A, computed both from Cotes' differential formula and from the accurate formula. They have been calculated for three different values of the angle C, namely, $C = 0.1$ radians, $C = 1$ radians, $C = 1.5$ radians. Two points are clear from the table:

(i)   The accuracy of the Cotes approximations varies with C;

(ii)   The accuracy increases, as C approaches 90°.

Reference will be made later to Theorem VIII.

Theorem VIII (Fig. 35)
In any plane triangle ABC, AB and AC being held constant.

$$\frac{d(B)}{d(BC)} = \frac{\cot C}{BC},$$

where the variations are small.

Proof
With reference to the diagram (Fig. 35)

$$\frac{HI}{ED} = \frac{HI}{EC} \times \frac{EC}{ED}, \; = \frac{r}{BC} \times \frac{\cot C}{r},$$

i.e.,

$$\frac{d(B)}{d(BC)} = \frac{\cot C}{BC},$$

as required. The spherical analogue of this theorem, namely Theorem XXIII, can be applied to the determination of solar mid-day by the method of corresponding altitudes, making allowance for the change in the sun's declination, as is shown later.

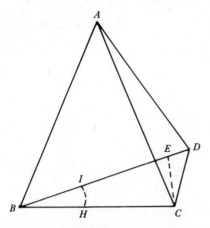

Fig. 35. Small variations in the angle *B* and the side *BC* are considered. From *Harmonia Mensurarum*, Cambridge University Press, 1722.

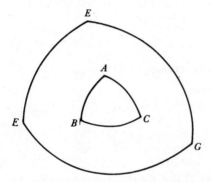

Fig. 36. *E*, *F* and *G* are the nearer poles of the arcs *BC*, *CA* and *AB* of the spherical triangle *ABC*. From *Harmonia Mensurarum*, Cambridge University Press, 1722.

The second and longer part of Aestimatio Errorum is taken up with results, similar to the foregoing and obtained by similar methods, for variations in the elements of spherical triangles. Cotes first establishes Lemma IV, as follows:

Lemma IV (Fig. 36)
ABC is a spherical triangle, *E*, *F*, *G*, are the nearer poles of *BC*, *CA*, *AB* respectively. $d(FG)$, $d(EG)$, $d(EF)$ = measure of $d(A)$, $d(B)$, $d(C)$ respectively, and the measures of $d(E)$, $d(F)$, $d(G)$ are equal to $d(BC)$, $d(CA)$, $d(AB)$ respectively.

These results follow from the fact that $EF = 180° - C$, and so on, and the lemma is useful for replacing results about differentials of

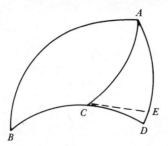

Fig. 37. Small variations in the arcs *BC* and *CA* of the spherical triangle *ABC* are considered. From *Harmonia Mensurarum*, Cambridge University Press, 1722.

sides and angles with results about differentials of angles and sides; rather like a principle of duality. I have listed the theorems in Fig. 41 and give below Cotes' proofs of Theorems XI, XII and XXIII, in brief notation.

Theorem XI (Fig. 37)
In spherical triangle ABC, B and AB being held constant,

$$\frac{d(BC)}{d(AC)} = \frac{r}{\cos C}.$$

Proof
Triangle *ABC* becomes triangle *ACD*. Put $AE = AC$. Then, if the variations are small, angles *ACE* and *CED* are right angles. Thus,

$$\frac{CD\ (=d(BC))}{DE\ (=d(AC))} = \frac{r}{\cos D} = \frac{r}{\cos C}$$

(*r* being the radius of the canon).

Theorem XII
With the same notation and conditions as in Theorem XI,

$$\frac{d(A)}{d(C)} = \frac{r}{\cos AC}.$$

Proof
In Lemma IV, let *G* and *EG* be held constant. Then, by Theorem XI,

$$\frac{d(FG)}{d(FE)} = \frac{r}{\cos F}$$

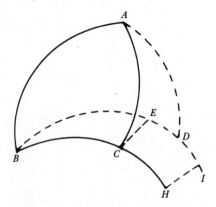

Fig. 38. Small variations in the angle *B* and the arc *BC* of the spherical triangle *ABC* are considered. From *Harmonia Mensurarum*, Cambridge University Press, 1722.

and hence, by Lemma IV,

$$\frac{d(A)}{d(C)} = \frac{r}{\cos AC},$$

as required.

Theorem XXIII (Fig. 38) (summarised)
In spherical triangle ABC, AB and AC are held constant, then

$$\frac{d(B)}{d(BC)} = \frac{\cot C}{\sin BC}.$$

Proof
Triangle *ABC* becomes triangle *ABD* and *AD* remains equal to *AC*. Let *BE* = *BC*. Let *BH, BI* be quadrantal arcs.

$$\frac{HI}{ED} = \frac{HI}{EC} \times \frac{EC}{ED},$$

hence this ratio is equal to

$$\frac{r}{\sin BC} \times \frac{\cot C}{r},$$

as required (*r* being the radius of the canon). Note that, *BH* and *BI* being quadrantal arcs, $HI = \delta(B)$. And *BE* being equal to *BC*, *ED* is $\delta(BC)$.

It is clear that Theorem XI is the spherical analogue of Theorem I and that Theorem XII is the dual of Theorem XI (in the sense in

Fig. 39. *Summary of the results for plane triangles in Aestimatio Errorum*

Theorem	Fixed quantities	Cotes' formula for small variations	Nature of the approximation involved	Delambre's correct results
I	AB, B	$\dfrac{d(BC)}{d(AC)} = \dfrac{r}{\cos C}$	Second order quantities, and δA, neglected	$d(AC) = \dfrac{\delta(BC)\cos C + 2AC\sin^2(\delta A)/2}{\cos(\delta A)}$
II	AB, B	$\dfrac{d(A)}{d(AC)} = \dfrac{\tan C}{AC}$	$\sin(\frac{1}{2}\delta A) = \frac{1}{2}\delta A$ on LHS and $\delta A = 0$ on RHS (see details below)	$\dfrac{2\sin\frac{1}{2}\delta A}{\delta(AC)} = \dfrac{\sin(C-\delta A)}{\cos(C-\frac{1}{2}\delta A)} \cdot \dfrac{1}{AC}$
III	AB, B	$\dfrac{d(C)}{d(DC)} = -\dfrac{\sin C}{AC}$	Second and higher powers of $\delta(BC)$ neglected. Sign not discussed	$\delta A = -\dfrac{\delta(BC)}{AC}\dfrac{\sin C}{\sin 1^0} + \dfrac{\delta(BC)^2}{AC}\dfrac{\sin 2C}{\sin 2^0}$ $- \dfrac{\delta(BC)^3}{AC}\dfrac{\sin 3E}{\sin 3^0} + \cdots$
IV	A, BC	$\dfrac{d(C)}{d(AB)} = \dfrac{\tan C}{AB}$	Neglect $\sin^2(\frac{1}{2}\delta C)$	$\delta(AB) = AB\cot C\sin\delta C - 2AB\sin^2(\frac{1}{2}\delta C)$
V	A, BC	$\dfrac{d(AB)}{d(AC)} = \dfrac{\cos C}{\cos B}$	Neglect $\frac{1}{2}\delta B, \frac{1}{2}\delta C$	$\dfrac{\delta(AB)}{\delta(AC)} = \dfrac{\sin\frac{1}{2}\delta C\cos(C+\frac{1}{2}\delta C)}{\sin\frac{1}{2}\delta B\cos(B+\frac{1}{2}\delta B)}$
VI	AB, AC	$\dfrac{d(A)}{d(B)} = \dfrac{BC}{AC} \cdot \dfrac{1}{\cos C}$	Neglect small quantities of first order on RHS (use sine rule on LHS)	$\dfrac{\sin\delta B}{\sin\delta A} = \dfrac{AC}{BC+\delta BC}\{\cos C + \sin C\tan\frac{1}{2}\delta A\}$
VII	AB, AC	$\dfrac{d(A)}{d(BC)} = \dfrac{\operatorname{cosec} C}{AC}$	$\sin(\frac{1}{2}\delta A) = \frac{1}{2}\delta A$, and neglect $\frac{1}{2}\delta(BC)$ (use $AB\sin A = BC\sin C$, $\frac{1}{2}\delta(A)$)	$\dfrac{2\sin(\frac{1}{2}\delta A)}{\delta(BC)} = \dfrac{BC+\frac{1}{2}\delta BC}{AB\cdot AC\sin(A+\frac{1}{2}\delta A)}$
VIII	AB, AC	$\dfrac{d(B)}{d(BC)} = \dfrac{\cot C}{BC}$	Neglect $\delta A, \delta B$ on RHS	$\dfrac{2BC\sin\frac{1}{2}\delta B}{\delta BC} = \dfrac{\cos(C-\frac{1}{2}\delta A+\delta B)}{\sin(C-\frac{1}{2}\delta A+\frac{1}{2}\delta B)} \cdot \dfrac{\tan B}{\tan C}$
IX	AB, AC	$\dfrac{d(C)}{d(B)} = \dfrac{\tan C}{\tan B}$	Neglect terms of second order	$\dfrac{\sin\delta B - 2\sin^2\frac{1}{2}\delta B\tan B}{\sin\delta C - 2\sin^2\frac{1}{2}\delta C\tan C} = \dfrac{\tan B}{\tan C}$
X	$A, B\ (\therefore C)$	$\dfrac{d(AB)}{AB} = \dfrac{d(BC)}{BC}$ $= \dfrac{d(CA)}{CA}$		The only accurate formula, i.e., not an approximation

Fig. 40. *Theorem II, table showing percentage variation in AC due to a given variation in angle A, computed (i) by Cotes' approximate formula (ii) the accurate formula, for C = 0.1, 1.0 and 1.5 radians*

Variation in A in minutes	% variation in AC Cotes'	Accurate	Variation in A in minutes	% variation in AC Cotes'	Accurate	Variation in A in minutes	% variation in AC Cotes'	Accurate
5	1.4500	1.4710	5	0.0093	0.0009	5	0.0103	0.0104
10	2.8992	2.9861	10	0.1868	0.1876	10	0.0206	0.0211
15	4.3488	4.5475	15	0.2802	0.2819	15	0.0309	0.0319
20	5.7984	6.1571	20	0.3736	0.3766	20	0.0413	0.0430
25	7.2479	7.8173	25	0.4669	0.4718	25	0.0516	0.0542
30	8.6975	9.5305	30	0.5603	0.5673	30	0.0619	0.0657
35	10.1471	11.2992	35	0.6537	0.6633	35	0.0722	0.0774
40	11.5967	13.1263	40	0.7471	0.7596	40	0.0819	0.0894
45	13.0463	15.0416	45	0.8405	0.8563	45	0.0928	0.1015
50	14.4959	16.9672	50	0.9339	0.9534	50	0.1031	0.1138
55	15.9455	18.9875	55	1.0273	1.0510	55	0.1134	0.1264
60	17.3951	21.0792	60	1.1207	1.1489	60	0.1238	0.1392
	C = 0.1 radian			C = 1 radian			C = 1.5 radians	

Fig. 41. *Summary of the results for spherical triangles in Aestimatio Errorum*

Theorem	Constant elements	Cotes' differential formula	Plane analogue	Notes	La Caille's remarks on accuracy
XI	B, AB	$\dfrac{d(BC)}{d(AC)} = \dfrac{1}{\cos C}$	Th. I		Exact if $B = 90°$ Less exact as B diminishes and $BC \to 90°$
XII	B, AB	$\dfrac{d(A)}{d(C)} = \dfrac{1}{\cos AC}$		Dual of XI	The smaller B is and the closer BC is to $90°$, the less exact is the result
XIII	B, AB	$\dfrac{d(A)}{d(AC)} = \dfrac{\tan C}{\sin AC}$	Th. II		Exact as in I
XIV	B, AB	$\dfrac{d(BC)}{d(C)} = \dfrac{\tan AC}{\sin C}$		Dual of XII	The smaller B is and the closer BC is to $90°$, the less exact
XV	B, AB	$\dfrac{d(A)}{d(BC)} = \dfrac{\sin C}{\sin AC}$	Th. III		If $B = 90°$ and $BC = 90°$, exact
XVI	B, AB	$\dfrac{d(C)}{d(AC)} = \dfrac{\tan C}{\tan AC}$		Can be obtained from XI and XIV	Always fairly good, except as B approaches $90°$
XVII	A, BC	$\dfrac{d(C)}{d(AB)} = \dfrac{\tan C}{\tan AB}$	Th. IV		More exact as AB, C approach $90°$ Less exact as AC, B approach $90°$
XVIII	A, BC	$\dfrac{d(AB)}{d(AC)} = \dfrac{\cos C}{\cos B}$	Th. V		Accuracy increases as BC increases
XIX	A, BC	$\dfrac{d(C)}{d(B)} = \dfrac{\cos AB}{\cos AC}$		Dual of XVIII	More exact as B increases

XX	A, BC	$\dfrac{d(AB)}{d(B)} = \dfrac{\tan AB}{\tan C} \cdot \dfrac{\cos AB}{\cos AC}$	Alternatively, $\dfrac{d(AB)}{d(B)} = \dfrac{\tan AC}{\tan B} \cdot \dfrac{\cos C}{\cos B}$ (result from the three previous theorems)	Both require differentials to be small
XXI	AB, AC	$\dfrac{d(A)}{d(B)} = \dfrac{\sin BC}{\sin AC} \cdot \dfrac{1}{\cos C}$	Th. VI	More exact as AC, BC approach 90° More exact as AB, BC increase beyond 90°
XXII	AB, AC	$\dfrac{d(A)}{d(BC)} = \dfrac{\operatorname{cosec} C}{\sin AC}$	Th. VII	More exact as AC, BC approach 90°
XXIII	AB, AC	$\dfrac{d(B)}{d(BC)} = \dfrac{\cot C}{\sin BC}$	Th. VIII	Both more exact as AC, BC approach 90°
XXIV	AB, AC	$\dfrac{d(C)}{d(B)} = \dfrac{\tan C}{\tan B}$	Th. IX	Always fairly exact
XXV	B, C	$\dfrac{d(BC)}{d(AC)} = \dfrac{\sin A}{\sin B} \cdot \dfrac{1}{\cos AB}$	Dual of XXI	More exact as B, C approach 90° More exact as AC approaches 90°
XXVI	B, C	$\dfrac{d(BC)}{d(A)} = \dfrac{\operatorname{cosec} AB}{\sin B}$	Dual of XXII	More exact as A, C approach 90°
XXVII	B, C	$\dfrac{d(AC)}{d(A)} = \dfrac{\cot AB}{\sin A}$	Dual of XXIII	More exact as B, C approach 90° More exact as A, C approach 90°
XXVIII	B, C	$\dfrac{d(AB)}{d(AC)} = \dfrac{\tan AB}{\tan AC}$	Dual of XXIV	Always exact so long as $d(AC)$ and $d(BC)$ can be supposed as straight lines

which I have used it) obtained by the use of Lemma IV. Theorem XXIII is the spherical analogue of Theorem VIII. In the table (Fig. 41) I have summarised the relationships for all the theorems and shown La Caille's reordering and comments on the best conditions for each theorem. From the table, Cotes' neat and efficient system, whereby all possible cases are considered, is apparent. (La Caille's reordering of the eighteen theorems into twenty-four seems a little prolix. After all, if in a triangle XYZ, angles X and Y are held constant, there seems little point in giving formulae for both $[d(YZ)]/[d(Z)]$ and $[d(ZX)]/[d(Y)]$.)

Cotes completes the treatise with two scholia. In Scholium I he says he has now shown all cases of variation among the elements of the triangles when two of the elements remain constant; however, there are many other ways of computing errors, but it would be too great a labour to outline them all. He will briefly explain the following method: Assume the error in some quantity A arises from errors in quantities B, C, D. Express A in terms of B, C and D, and hence compute the value of their fluxions by Newton's method. Then the error in A will be given by (the sum of) the errors in B, C, and D, signs properly chosen, each multiplied by their fluxions, i.e., Cotes is here giving the formula for total differential

$$d(A) = d(B)\dot{B} + d(C)\dot{C} + d(D)\dot{D}.$$

He goes on to say that, in trigonometrical expressions, it is not always easy to bring A, B, C, D together into one expression, but that sines, tangents and secants can often be substituted for the fluxions of sines, tangents and secants when the variations are small, and these quantities are then given by the fluxion of the arc, as in the first three lemmas.

In Scholium II Cotes pursues further his thoughts about errors. First, a simple illustrative example, with plane triangles. Suppose the length of AC in the triangle ABC, right-angled at A is given. (Fig. 42). Then AB can be calculated if $A\hat{C}B$ is known. The error in AB arising from an error in $A\hat{C}B$ is given by

$$\frac{d(AB)}{d(A\hat{C}B)} = \frac{BC}{\sin(A\hat{B}C)},$$

by Theorem III,

$$= \frac{2AB}{\sin 2(A\hat{C}B)}$$

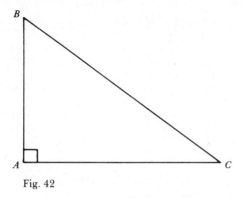

Fig. 42

(because $AB = BC \sin A\hat{C}B$). Hence,

$$\frac{d(AB)}{AB} = \frac{2d(A\hat{C}B)}{\sin 2A\hat{C}B},$$

and hence the error in AB is least when $\sin 2A\hat{C}B$ is greatest, i.e., when $A\hat{C}B = 45°$. Suppose then, says Cotes, that the length of AB is to be determined by instrumental measurement of angle C, then the most reliable value of AB will be obtained when $A\hat{C}B$ is 45°, instrumental errors then having the smallest effect. Thus, if the radius has 10 000 000 parts, the error in AB arising from an error of one first minute in measuring $A\hat{C}B$ will be 2909 parts ($2 \times \pi \times 10^7 \div (360 \times 60)$). Hence,

$$\frac{d(AB)}{AB} = \frac{2 \times 2909}{10^7} = \frac{1}{1719}.$$

The proportional error in AB will increase or decrease as the error increases or decreases, and in the same proportion. If $A\hat{C}B$ is not 45°, the error will increase in the proportion of the radius to the sine of twice the angle. These results (Cotes does not give the working in detail) serve to emphasise the point made both by La Caille and Delambre, that care is needed in the use of the formulae, and that unthinking use of a formula not suitable for a particular observation could lead to errors which are not negligible.

Next, an example applied to spherical triangles, thus. P is the pole, V the observer's vertex, and S the position of a star (Fig. 43): 'A common problem in astronomy' [says Cotes] 'is the determination of sidereal time by observing the hour-angle *VPS*.' Cotes proposes to compute the error arising in the time, from a given error of observation of the star's altitude. *PV* and *PS* being given, we have, from

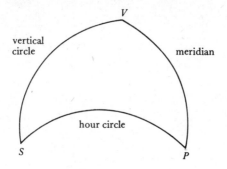

Fig. 43. Theorem XXII is applied to the spherical triangle *VPS*. From *Harmonia Mensurarum*, Cambridge University Press, 1722.

Theorem XXII,

$$d(V\hat{P}S) = \frac{d(VS)}{\sin(P\hat{V}S)\sin(PV)},$$

i.e.,

$$\text{error in hour-angle} = \frac{\text{error in altitude of star}}{\sin P\hat{V}S \sin(90° - \text{altitude of pole})}.$$

The error will clearly be least when the star is selected so that $P\hat{V}S$ (the angle between the vertical circle through the star, and the observer's meridian) is 90°, and when *VP* is 90°, i.e., when the observer is at the equator. Cotes now states the following results:

(i)   At the equator, an error of one first minute in the observation of the star's altitude will result in an error of four seconds in the hour-angle;

(ii)   In latitude 45°, an error of $5\frac{2}{3}$ seconds will result;

(iii) In latitude 50° the error will be $6\frac{2}{9}$ seconds;

(iv)   In latitude 55°, the error will be $6\frac{37}{38}$ seconds.

(We can quickly check these results. Thus

(i) $d(V\hat{P}S) = 1/360 \times 60$ of a day $= 1/21\,600$ of a day $= 4$ seconds;

(ii)   error in time $= 1/21\,600(1/\sin 45°)$ of a day $= 5.656$ seconds;

(iii) and (iv) follow similarly.)

All these results depend on choosing a star in a vertical circle at right angles to the meridian. If some other vertical circle is chosen, Cotes remarks that the error in the time will increase as the radius to the sine of the angle *PVS*. Similarly, if the error in measuring the altitude is greater or less than one minute, the error in the time will be increased or decreased, and in the same proportion.

Referring to the Clare paper mentioned previously (p. 90), we can now see that what Cotes is describing is the same application of his Theorem XXIII as that just discussed. The error in time arising from an error of one degree – clearly a copying error for one minute – in the star's altitude is given as 6′ 32″ in the latitude of Cambridge. If we apply the theorem we find error in time

$$= \frac{24 \times 60 \times 60}{360 \times 60 \times \cos 52° 15'} = 6.533628' = 6' \, 32'',$$

as given. An error of 1 second in time would arise from an error in altitude of

$$\frac{1}{6.533628} = 0.153' = 9'',$$

again as given.

Aestimatio Errorum concludes with the interesting statement that when a number of slightly different observations are made of a particular quantity, *the most probable value* of the correct observation can be found as follows:

Sit *p* locus Objecti alicujus ex Observatione prima definitus, *q, r, s,* ejusdem objecti locae Observationibus subsequentibus; sint insuper *P, Q, R, S* pondera reciproce proportionalia spatiis evagationem, per quae se diffundere possint errores ex Observationibus singulis prodeuntes, quaeque dantur ex datis Errorum Limitibus; & ad puncta *p, q, r, s* posita intelligantur pondera *P, Q, R, S.* & inveniatur eorum gravitatis centrum *Z*: dico punctum *Z* fore Locum Objecti maxime probabilem, qui pro vero ejus loco tutissime haberi potest.

Let *p* be the place of some object defined by observation, *q, r, s* the places of the same object from subsequent observations. Let there also be weights *P, Q, R, S* reciprocally proportional to the displacements arising from the errors in the single observations, and which are given by the limits of the given errors; and the weights *P, Q, R, S* are conceived as being placed at *p, q, r, s,* and their centre of gravity *Z* is found: I say the point *Z* is the most probable place of the object [Fig. 44].

(R. Cotes in Opera Miscellanea, *Harmonia Mensurarum.*)

It has been claimed that this method of Cotes' for determining the most probable value, from a number of readings, each subject to small errors, is equivalent to the method of least squares (used by Legendre in 1806). This claim is discussed by J. B. J. Delambre in *Histoire de l'Astronomie au Dixhuitième Siècle,* p. 455 and by D. T. Whiteside, in *The Mathematical Papers of Isaac Newton,* vol. 6 (8 vols., Cambridge, 1974) p. 51, note. The following is based on the remarks of these two authorities, somewhat amplified.

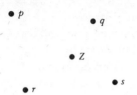

Fig. 44. From *Harmonia Mensurarum*, Cambridge University Press, 1722.

Suppose, relative to some suitable set of Cartesian axes,

$$Z = (x_1, x_2, x_3,); \qquad p = (p_1, p_2, p_3,); \qquad q = (q_1, q_2, q_3,);$$

$$r = (r_1, r_2, r_3,); \qquad s = (s_1, s_2, s_3,).$$

Since the weights are to be reciprocally proportional to the errors, denote these errors by

$$e_p, e_q, e_r, e_s,$$

and then

$$P = \frac{k}{e_p}, \qquad Q = \frac{k}{e_q}, \qquad R = \frac{k}{e_r}, \qquad S = \frac{k}{e_s},$$

where k is some constant of proportionality. Then putting $P + Q + R + S = W$ and taking moments about the coordinate axes, we have the following sets of equations.

$$Wx_i = Pp_i + Qq_i + Rr_i + Ss_i \quad (i = 1, 2, 3). \tag{1}$$

Also

$$e_p^2 = (x_i - p_i)^2, \qquad e_q^2 = (x_i - q_i)^2 \quad (i = 1, 2, 3)$$
$$e_r^2 = (x_i - r_i)^2, \qquad e_s^2 = (x_i - s_i)^2 \quad (i = 1, 2, 3). \tag{2}$$

If we put T for $e_p^2 + e_q^2 + e_r^2 + e_s^2$ and require T to be a minimum, we must have

$$\frac{\partial T}{\partial x_i} = 0 \quad (i = 1, 2, 3).$$

Hence,

$$4x_i = p_i + q_i + r_i + s_i \quad (i = 1, 2, 3).$$

This is the result obtained by putting $P = Q = R = S$ in (1); and hence the position of $Z(x_1, x_2, x_3)$ is known. Thus, if P, Q, R, S are equal weights, the result is that obtained by applying the method of least squares.

Cotes' proposal certainly leads to a system of equations which, with some difficulty if the observations are at all numerous, is solvable for x_1, x_2, x_3 and the errors e_p, e_q, e_r, e_s. His concept of attaching weights to the errors, these weights being related to the errors by a stated law (in this case a simple reciprocal law), and then finding a mean of the weighted errors, can fairly be regarded as a serious attempt at error analysis. It is clearly more than the finding of a convenient mean reading; and Cotes himself states that the result is the most *probable* reading.

In his *Doctrine and Applications of Fluxions*, vol. I (2 vols., London, 1750), Thomas Simpson treated the subject of small variations in the elements of spherical triangles, much on the lines of Cotes, but without reference to him. Simpson was accused of plagiarism in *The Monthly Review* (December 1740), p. 130. The writer, signing himself Cantabrigiensis, says of Simpson's Section I, Part II, Theorems I, II and III: 'This is all taken from Mr Cotes' Aestimatio Errorum, where it is elegantly demonstrated.'

Among the Lucasian papers in the University Library at Cambridge (Add MS 3960, fos. 5, 67, 68) is a paper which, according to D. T. Whiteside (*The Mathematical Papers of Isaac Newton*, vol. 4 (8 vols., Cambridge, 1971), p. 421 is in Horsley's handwriting. The paper consists of solutions of problems on the variations of the sides of a right-angled triangle, the hypotenuse being held constant, the problems, says Whiteside, being taken from a paper of Newton's on 'The Geometry of Curved Lines'. For example, Proposition I of this paper proves in a right-angled triangle, right-angled at A,

$$\frac{\overset{\cdot}{\overline{AB}}}{\overset{\cdot}{\overline{AC}}} = \frac{AB}{AC}.$$

The Propositions are labelled I, II, III, IV, VII, VIII, and the only point of mild interest here is that the demonstrations of Propositions II and IV are not given, being simply referred by the author to Cotes' Aestimatio Errorum, Propositions IV and I respectively.

I conclude with two examples of applications of the theorems in Aestimatio Errorum.

(i)   *To find the apparent motion of the stars caused by the procession of the equinoxes* (see N. L. de La Caille, *Histoires de l'Académie Royale des Sciences*). P is the pole of the equator (Fig. 45); p is the pole of the ecliptic; E is the position of a star; Pp and E are constant. The longitude of the star, i.e., the complement of $\widehat{BpP}$, is variable by $50''$

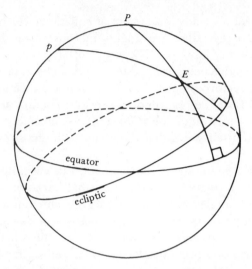

Fig. 45. The precession of the equinoxes, represented by small changes in the angle *pPE*, causes small changes in the astronomical coordinates of the star *E*. From *Harmonia Mensurarum*, Cambridge University Press, 1722.

per year. By Formula XIII (i.e., Cotes Form XXI)

$$\frac{d(p)}{d(P)} = \frac{R \sin PE}{\sin PE \times \cos E}\left(= \frac{\sin p}{\sin P} \times \frac{1}{\cos E} = \frac{\sin PE \times \tan E}{\sin P \times \sin Pp}\right),$$

from sine rule (La Caille gives only triangle *pPE*), and from Formula XV (i.e., Cotes Form XXII)

$$\frac{d(p)}{d(pE)} = \frac{1}{\sin PE \times \sin E}\left(= \frac{1}{\sin Pp \times \sin P}\right).$$

Since $d(p)$ is known to be 50″ annually, $d(P)$ and $d(pE)$ can be found.
(ii)   *To apply Cotes' Theorem XXIII to the problem of the determination of solar mid-day by the method of corresponding altitudes, making allowance for the change in the sun's declination* (the following is based on the description given by Robert Smith in his *Compleat System of Opticks in Four Books*, vol. I). In the diagram (Fig. 46) *P* is the pole of the celestial sphere, *V* the observer's vertex, *A* and *C* the position of the sun at the morning and evening observations. *ABCD* is a circle of constant altitude. *ACE* is a circle of constant declination. The sun does not pass through *C* because of its change in declination, but through some point such as *D*. The problem is to determine the change in the hour-angle *CPV* consequent upon the change in *CP*, namely *DE*; i.e., we require to calculate $d(P)$ when triangle *PCV* becomes triangle

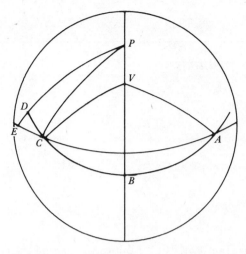

Fig. 46. Solar mid-day occurs midway between the times of observation of the sun at *A* and at *E*, two points with corresponding altitudes.

PDV. From Theorem XXIII, we have

$$\frac{d(P)}{d(PC)} = \frac{\cot PCV}{\sin PC}, \quad \text{i.e.,} \quad \frac{DC}{DE} = \frac{\cot PCV}{\sin PC}.$$

Since the change in declination can be observed (*PD* is the complement of the declination), *DE* is known and hence the change in hour-angle caused by this change in declination is known. Thus the time interval between the two equal (or corresponding) altitude observations, at *A* and *E*, is known and solar mid-day can be determined.

6

Numerical methods

Principia, first edition (London, 1687), Book III, Lemma V: *To find a curved line of the parabolic kind which shall pass through any given number of points*, was the first publication to the world of Newton's method of divided differences, for fitting a polynomial to a number of tabulated points. In this lemma, Newton presented the interpolation formula which bears his (and Gregory's) name and extended it to the case of unequal tabular intervals, using divided differences. Students of *Principia* will know that the immediate application of the method was, in Lemma VI: *Certain observed places of a comet being given, to find the place of the same at any intermediate given time.*

The motion of comets was the source of some of the most effective anti-Cartesian arguments, and in his work for *Principia*, Cotes was very much involved in computations to determine cometary orbits. Newton's comment in his own very brief Preface to the second edition, 'and the theory of the comets was confirmed by more examples of the calculation of their orbits, done also with greater accuracy', reflects the importance he attached to this work. The demonstration and applications of the method form the substance of the first of Cotes' two papers on what is now loosely termed 'numerical methods'. The papers deal with interpolation procedures, approximate integration, and the construction of tables. The first, called De Methodo Differentiali Newtoniana, was written in 1708 and was included in Cotes' public lectures in 1709.

In 1711, William Jones collected together some small treatises of Newton's, including his *Methodo Differentiali* (London, 1711) published them in a single volume, and sent Cotes a copy. The following extracts from the ensuing correspondence reveal something of the background:

You have highly obliged the Mathematical part of ye World by collecting into one Volume those curious and usefull Treatises which were before too much dispersed but more especially by ye publication of ye Analysis per Aequationes infinitas & the Methodus Differentialis .... The inclosed Paper is what I wrote about 3 Yeares ago & read to my Auditors in our Schools in 1709. I have sent it to You as it relates to the Methodus Differentialis but more particularly as a small acknowledgment of my gratitude for having received y^t and the other excellent Treatises from Your hands & as a token of my hearty freind-ship & sincere good will to You.

Not having heard any thing of ye book till I saw it I received it with ye additional pleasure of a Surprize.

(Cotes to Jones, 15 February 1711 [4], Letter CII.)

The paper concerning S^r Is. Newton's method of Interpolation, which you have bin pleas'd to send me, being done so very neat, that it wou'd be an injury to the Curious in these Things, to be kept any longer without it; therefore must desire you'd grant me leave to publish it in the *Phil. Trans.* you may be assur'd, that I don't move this to you, without Sr Isaac's approbation, who I find is no less willing to have it done.

(Jones to Cotes, 17 September 1711 [4], Letter CIII.)

I thank You for Your kind offer of recommending my Paper to the Publick; but I am of opinion it is not of so great use as to deserve to be printed after Sr Isaac's Methodus Differentialis

(Cotes to Jones, 30 September 1711 [4], Letter CIV.)

First then to Cotes' paper, De Methodo Differentiali Newtoniana, and the tail piece which he added having seen Newton's treatise. It is quite a short paper, consisting of an introduction, seven propositions, a Scholium and a postscript. In the Introduction, Cotes refers to the *Principia* results mentioned at the beginning of this chapter, and says that he will demonstrate (i.e., prove) Newton's results, which had not so far been done, and add some things of his own. Proposition I is a preliminary result which Cotes needs in order to prove that the nth divided difference of a polynomial of degree n, is constant. The Proposition is as follows: *Any series of quantities having been proposed, from all except the last, form the sum of all possible terms of a given dimension. From all except the first form the sum of all possible terms of the same dimension. Subtract the latter from the former and divide the result by the first term minus the last term. The quotient will be the sum of all possible terms of the next lower dimension.* Cotes does not give a general proof, but infers the truth of the result from a number of examples, of which two will suffice:

Example 2

Terms	Dimension	First sum	Second sum	Difference	Quotient on division by $a-e$
a, b, c	3	$a^3 + a^2b$ $+ ab^2 + b^3$	$b^3 + b^2c$ $+ bc^2 + c^3$	$a^3 + a^2b + ab^2$ $- b^2c - bc^2 - c^3$	$a^2 + b^2 + c^2 + ab$ $+ bc + ca$

It can be seen that the result is the sum of all terms of degree 2 formed from a, b, and c.

Example 4

Terms	Dimension	First sum	Second sum	Difference	Quotient on division by $a-e$
a, b, c, d, e	1	$a+b+c+d$	$b+c+d+e$	$a-e$	1

The last result clearly does not accord with the general statement, but has to be regarded as a special case, and is needed to prove the key Proposition II: *the nth divided differences of a polynomial $y = f + gx + hx^2 + kx^3 + \cdots$ of degree n are constant and equal to the coefficient of x^n*

Cotes proceeds as follows

$$
\begin{array}{c|c|c||c|c||c|c||c|c||c|c}
 & & & a & \alpha & & & & & & \\
 & \mu' & \lambda' & b & \beta & r' & \rho' & s' & \sigma' & t' & \tau' \\
\nu' & \mu'' & \lambda'' & c & \gamma & r'' & \rho'' & s'' & \sigma'' & t'' & \tau'' \\
\nu'' & \mu''' & \lambda''' & d & \delta & r''' & \rho''' & s''' & \sigma''' & & \\
 & & \lambda'''' & e & \varepsilon & r'''' & \rho'''' & & & & \\
\end{array}
$$

Let $x = a, b, c, d, e, \ldots$ and the corresponding values of $y = \alpha, \beta, \gamma, \delta, \varepsilon$. Then

$$\lambda' = a - b, \qquad \lambda'' = b - c, \text{ etc.,}$$

$$\mu' = a - c, \qquad \mu'' = b - d, \text{ etc.,}$$

$$\nu' = a - d, \qquad \nu'' = b - e, \text{ etc.,}$$

and these are the divisors to be used at each stage in forming the divided differences. ρ', σ', τ', etc., are the divided differences, thus

$$r' = \alpha - \beta \qquad \rho' = \frac{\alpha - \beta}{\lambda'}$$

$$s' = \rho' - \rho'' \qquad \sigma' = \frac{\rho' - \rho''}{\mu'}$$

$$r'' = \beta - \gamma, \qquad \rho'' = \frac{\beta - \gamma}{\lambda''}.$$

Cotes sets out the whole scheme in detail.

In Case 1, if $y = x^n$ we have $\rho' = (a^n - b^n)/(a - b), \rho'' = (b^n - c^n)/(b - c)$, i.e., the ρ are the sum of all the terms of degree $n - 1$ formed from a, b; from b, c; from c, d; & Co. Similarly the σ are the sum of all the terms of degree $n - 2$ formed from a, b, c; b, c, d; c, d, e & Co. Continuing the process, clearly 1 is reached as in Proposition I Example 4. This will be at the first difference if $y = x$, at the second difference if $y = x^2$, and so on.

In Case 2, if $y = kx^n$, the constant difference will be k, not 1.

In Case 3, if $y = g + hx + kx^2 + \cdots$. Clearly (says Cotes), the differences arising from the lowest terms will vanish first, then those from the next highest and so on, until the nth divided difference, which will be the coefficient of the highest term.

In Proposition III, this result is used to derive an interpolation formula, equivalent to that given by Newton in *Principia* (previously referred to). Cotes modifies the above table, and writes

		λ	Z	P	r	ρ	s	σ	t	τ
	μ	λ'	a	α		ρ				
ν	μ'	λ''	b	β	r'	ρ'	s'	σ'		
ν'	μ''	λ'''	c	γ	r''	ρ''	s''	σ''	t'	τ'
ν''	μ'''	λ''''	d	δ	r'''	ρ'''	s'''	σ'''	t''	τ''
			e	ε	r''''	ρ''''				

,

where the value P of y is required when $x = Z$. Assuming the τ, τ', τ'' to be the constant divided difference, τ is known, hence all the entries in the top line can be calculated (for example, $t = \tau\nu$, $\sigma = \sigma' + \tau\nu, \ldots$) and we have eventually,

$$P = \alpha + \rho'\lambda + \sigma'\lambda\mu + \tau'\lambda\mu\nu + \cdots.$$

This is the interpolation formula, equivalent to what is often referred to as Newton's first interpolation formula, which, in a convenient modern notation, is written

$$f(x) = f(0) + (x - x_0)[0\ 1] + (x - x_0)(x - x_1)[0\ 1\ 2]$$

$$+ \cdots (x - x_0)(x - x_1) \cdots (x - x_{n-1})[0\ 1\ 2, \ldots, n]$$

where

$$[0, 1] = \frac{f(x_0) - f(x_1)}{x_0 - x_1} = \frac{f(x_0)}{x_0 - x_1} + \frac{f(x_1)}{x_1 - x_0},$$

$$[0, 1, 2] = [0\ 1] - [1\ 2] = \frac{f(x_0)}{(x_0 - x_1)(x_0 - x_2)} + \frac{f(x_1)}{(x_1 - x_0)(x_1 - x_2)}$$

$$+ \frac{f(x_2)}{(x_2 - x_0)(x_2 - x_1)}.$$

As an illustration I give here the Cotes divided difference table for $f(x) = x^3$, tabulated at $x = 3, 4, 6, 10, 11$; the value of $f(x)$ at $x = 2$ being required.

			x	P	r	ρ	s	σ	t	τ
		−1	2							
	−2		3	27						
−4		−1	4	64						
−7	−3		6	216	−37	37				
−7		−2	10	1000	−152	76	−39	13		
	−6		11	1331	−784	196	−120	20	−7	1
		−4			−331	331	−135	27	−7	1
	−5									
		−1								

Using the Cotes formula

$$f(2) = P = \alpha + \rho'\lambda + \sigma'\lambda\mu + \tau'\lambda\mu\nu$$

$$= 27 + 37(-1) + 13(-1)(-2) + 1(-1)(-2)(-4)$$

$$= 8,$$

as required. Once again, we can observe that Cotes has taken a known result, and developed a thoroughly neat and usable form. It is worth remarking that the method is independent of the size of the tabular intervals, and indeed of the order in which they are tabulated.

In Proposition IV Cotes applies the method to determine a polynomial to fit given data. Thus, assuming $y = f + gx + hx^2 + kx^3$ to be the polynomial, but Z (i.e., x) $= 0$ in Proposition III and P($=y$) is f. As f is known, form $(y - f)/x$ at each tabular point,

$$\frac{y - f}{x} = g + hx + kx^2 + \cdots$$

and g can be found in the same way as f. Repeat the process, and each coefficient can then be found. Cotes does not discuss the accuracy or limitations of this method, the problems arising if the interpolated or extrapolated value of Z is far from the tabular point and so on. However, he clearly feels that more is needed, and suggests the

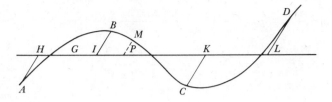

Fig. 47

following alternative procedure. Tabulate until the nth divided differences are equal. The value of the nth divided differences is the coefficient of the highest power, as has been shown. For example, if $y = f + gx + hx^2 + kx^3$, k is found. Form $y - kx^3 = f + gx + hx^2$, tabulate again, and h is found. Hence all the coefficients can be found.

Cotes now feels he has sufficiently established the basis for Newton's Lemma V – *To pass a parabolic curve through any given number of points* – as indeed he has. First however, in Proposition 7, he observes that an equation of the form $y = f + gx + hx^2 + kx^3 + \cdots$ of degree $n - 1$ can always be found to be satisfied by n values of x and y in one-to-one correspondence (*singularum ad singulus relatio*) – since n linear equations in the n coefficient result, and these can always be solved. Thus, Proposition VI is: *To find the curved line of the parabolic form, which passes through any given number of points.*

With reference to the diagram (Fig. 47), GH, GI, GK, GL (values of x) are related to AH, IB, KC, LD (values of y) as in Proposition V, hence the equation can be found. Alternatively, the equation can be found as in Proposition IV. The value MP corresponding to some other value GP of x can be found as in Proposition III.

Cotes adds a Corollary: *Hence it is possible to interpolate values.* This is the first time Cotes has used the word, (*interpolari*), and he chooses to illustrate the process by Newton's derivation of the expansion of $(1 + Q)^{m/n}$ from the table of $(1 + Q)^r$, r integral, in his own typically clear and concise manner, thus: With the notation of Proposition III, consider

$$
\begin{array}{c|l|c|c|c}
 & \dfrac{m}{n} \quad (1+Q)^{m/n} & & & \\
\lambda & & & & \\
\mu & 0 \quad 1 & & & \\
\nu & & & & \\
 & 1 \quad 1 + Q & \rho' & & \\
 & 2 \quad 1 + 2Q + Q^2 & & \sigma' & \\
 & 3 \quad 1 + 3Q + 3Q^2 + Q^3 & & & \tau' \\
\end{array}
$$

$$\lambda = \frac{m}{n}, \qquad \mu = \frac{m-n}{n}, \qquad \nu = \frac{m-2n}{n},$$

$$\rho' = Q, \qquad \sigma' = \frac{Q^2}{1 \times 2}, \qquad \tau' = \frac{Q^3}{1 \times 2 \times 3},$$

and

$$P = \alpha + \rho'\lambda + \sigma'\lambda\mu + \tau'\lambda\mu\nu$$

gives

$$(1+Q)^{m/n} = 1 + \frac{m}{n}Q + \frac{m}{n} \times \frac{m-n}{2n}Q^2 + \frac{m}{n} \times \frac{(m-n)}{2n} \times \frac{(m-2n)}{3n}Q^3 + \cdots.$$

Finally, to Proposition VIII which is: *To determine the areas of all curves, approximately.* Cotes first points out that through any number of given points on a given curve, it is possible to pass a parabolic curve. This will approximate to the given curve. Thus if $y = f + gx + hx^2 + kx^3 + \cdots$ is a parabolic curve passing through $ABCD$, the area $AKPM$, where K is the origin, and $KP = x$ will be

$$fx + \tfrac{1}{2}gx^2 + \tfrac{1}{3}hx^3 + \tfrac{1}{4}kx^4 + \cdots$$

approximately. There follows a lengthy and intriguing geometrical justification of this last result, which I shall omit from the present work.

Finally, there is the postscript which Cotes wrote, having seen Newton's treatise. In this, following suggestions made by Newton, Cotes developed what became known as the Cotes–Newton formulae for approximate quadrature. These are derived by considering the area under a parabola through three points at equal intervals on a given curve (Simpson's rule), a cubic through four such points (the three-eighths rule), and so on up to a curve of order ten through eleven equally-spaced points. The list is shown in Fig. 48. In these formulae, A is the sum of the first and last ordinates, B is the sum of the second and penultimate, C is the sum of the third and antepenultimate ordinates, and so on in alphabetical order. R is the length of the base (the notation is Newton's).

In his treatise, Newton gave an interpolation formula different from that used in *Principia*, and used by Cotes. Newton stated in a scholium that his propositions were useful for constructing tables, and for the solution of problems which depended on the quadrature of curves. As an example of the latter, Newton gave, without proof, the area under a parabolic curve between the first and last of four equally-spaced ordinates as $[(A + 3B)/8]R$, the first formula in Fig. 48. Cotes followed up both of these suggestions of Newton's with his usual

Numerus Ordinatarum.	Areæ
III.	$\dfrac{A + 4B}{6}R$
IV.	$\dfrac{A + 3B}{8}R$
V.	$\dfrac{7A + 32B + 12C}{90}R$
VI.	$\dfrac{19A + 75B + 50C}{288}R$
VII.	$\dfrac{41A + 216B + 27C + 272D}{840}R$
VIII.	$\dfrac{751A + 3577B + 1323C + 2989D}{17280}R$
IX.	$\dfrac{989A + 5888B - 928C + 10496D - 4540E}{28350}R$
X.	$\dfrac{2857A + 15741B + 1080C + 19344D + 5778E}{89600}R$
XI.	$\dfrac{16067A + 106300B - 48525C + 272400D - 260550E + 427368F}{598752}R$
&c.	

Fig. 48. Cotes' formulae for approximate quadrature. From *Harmonia Mensurarum*, Cambridge University Press, 1722.

thoroughness. The construction of tables is the substance of the next paper to be discussed, Canonotechnia, and the result $[(A + 3B)/8]R$ (a rule described by Cotes as *Pulcherrima et utilissima*), was the inspiration for the whole set of formulae in Fig. 48.

Cotes gives no hint of his method of derivation of these formulae. However, since the one example which Newton gave was clearly intended to illustrate an application of his second interpolation formula, it is reasonable to suppose that Cotes derived the formulae for approximate quadrature from this second interpolation formula. This is achieved by integration between suitable limits. As an illustration of Cotes' probable method, I derive the result for five equally-spaced ordinates. In the following, y_i $(i = 1-5)$ are the values of the ordinates at x_i $(i = 1-5)$. As with Newton, the tabular interval is taken as unity. The differences in the table are undivided, the divisors $2, 6, 24, \ldots$, appearing as the denominators in Newton's formula, namely,

$$y = k + xl + \frac{x^2}{2}m + \frac{x^3 - x}{6}m + \frac{x^4 - x^2}{24}n + \cdots$$

(where the k, l, m, n, ... are the entries in the central row of the difference table, or the mean of two differences).

y_0

$\quad y_0 - y_1$

$y_1 \qquad\qquad y_0 - 2y_1 + y_2$

$\quad y_1 - y_2 \qquad\qquad y_0 - 3y_1 + 3y_2 - y_3$

$y_2 \qquad\qquad y_1 - 2y_2 + y_3 \qquad\qquad\qquad y_0 - 4y_1 + 6y_2 - 4y_3 + y_4$

$\quad y_2 - y_3 \qquad\qquad y_1 - 3y_2 + 3y_3 - y_4$

$y_3 \qquad\qquad y_2 - 2y_3 + y_4$

$\quad y_3 - y_4$

y_4

If the limits of integration are taken as -2 and 2, it is clear that the second and fourth terms in the interpolation formula can be ignored, and we have

$$\text{Area} = \left\{ y_2 x + \frac{y_1 - 2y_2 + y_3}{2} \frac{x^3}{3} + \frac{y_0 - 4y_1 + 6y_2 - 4y_3 + y_4}{24}\left(\frac{x^5}{5} - \frac{x^3}{3}\right) \right\}_{-2}^{2},$$

which reduces to

$$4\left\{ \frac{7(y_0 + y_4) + 32(y_1 + y_3) + 12 y_4}{90} \right\}$$

$$= \frac{7A + 32B + 12C}{90} R \text{ in Cotes–Newton notation.}$$

Clearly, the complexity increases rapidly as the number of ordinates increases. Newton's Case II, for an even number of ordinates, requires slightly more work. The more general formulae for unequal tabular intervals, which Newton gave, were used by Cotes in his work on the construction of tables, Canonotechnia.

The Cotes formulae for approximate quadrature, again illustrate his skill in formulating useful rules and techniques, particularly in computational methods, and the relatively small effect which his work seems to have had. W. W. Johnson has given an excellent history of these formulae in an article called 'On Cotesian Numbers, Their History, Computation and values to $n = 20$', in the *Quarterly Journal of Pure and Applied Mathematics*, 46 (1915), 52–65 (see Fig. 49). Reference should also be made to A. Kowalewski, *Newton, Cotes, Gauss, Jacobi. Vier Grundlegende Abhandlungen uber Interpolation und genaherte Quadratur* (Leipzig, 1917).

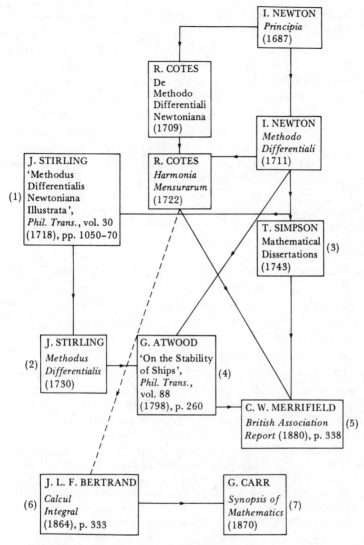

Fig. 49. The diagram summarises the details concerning Cotes–Newton approximate integration formulae, given by Professor W. W. Johnson.

Notes on the diagram (Fig. 49)

In the following, n is the number of intervals, $n + 1$ being the number of ordinates.

(1) In the *Philosophical Transactions of the Royal Society* paper, Stirling gives the results for even values of n, up to $n = 10$. He does not refer to Cotes.

(2) In his *Methodus Differentialis*, Stirling omitted the value for $n = 10$.

(3) Thomas Simpson gave results as far as $n = 6$. He does not mention Cotes, although he was familiar with Cotes' property of the circle.

(4) Atwood quotes Stirling's results up to $n = 8$ and supplies the results for odd values of n. He makes no reference to Cotes.

(5) Merrifield's report is the earliest account Johnson has found on Cotes' numerical work, and I have found none earlier.

(6) Bertrand's chapter is headed, 'Méthode d'Interpolation de Cotes', and gives results up to $n = 10$. There is nothing to suggest that Bertrand was familiar with Cotes' actual work: he gives the coefficients in a different form.

(7) Carr corrected Bertrand's work, but does not say that Cotes had done the work correctly.

CANONOTECHNIA

As we have seen, Cotes' reading of Newton's *Methodus Differentialis* (London, 1711), particularly Proposition III, Newton's second interpolation formula, was the starting point for Cotes' further work in this field. The following further extracts from the Cotes–Jones correspondence, most but not all of which is given in [4], show something of the background to this development.

The reading of that treatise gave me occasion to consider the subject again. My design related cheifly to the construction of tables & I found & examined a great variety of Methods which might be imploy'd for that purpose, but at length I fix'd upon one which I judg'd to be much the simplest and most convenient of any of those which offer'd themselves to my thoughts. The account of this method may make a second part to be annex'd to the paper I formerly sent you if you and your Mathematical Freinds shall think it of any use to be publish'd. I have not yet drawn it up in writing otherwise it might have accompanied this letter. The best definition I can give you in few Words is this; That in some respects it resembles Sr Isaac's Method in Prop. III of his *Methodus Differentialis*, but is in the main more nearly related to ye method of Mr Briggs defin'd in Cap. 13 of his *Arithmetica Logarithmica* Edit. Lond. Dr Gregory lib:v. Prop. XXV of his *Astronomy* refers his reader to a book of Gabriel Mouton *de Observationibus Diametrorum Solis & Lunae apparentium* [Lyons, 1670]. Though I do not much rely upon the D^{rs} recommendation, yet I would be glad to see the book or have some short account of it. If you could assist me in this respect I should be much obliged to you. I have formerly sent to my Bookseller at London, but he could not procure it.

(Cotes to Jones, 11 September 1710, Trinity College Library MS R 16 38, fo.
299 and also partly in [4], Letter CIV.)

And as you have some further Considerations about the Doctrine of Differences, I am assured, they cannot but be valuable; and if a few Instances of the application were given, perhaps it wou'd n't be amiss. Having tarried for some time for a convenient opportunity, I was at last oblig'd to send you Mouton's Book by the Carrier; tho it will only satisfy you that Dr Gregory had but a very Slender notion of the design, extent, & use of Lem. 5. Lib. 3 of the Principia; I hope it will not be long before you find leasure to send us what you have further done in this curious subject; no excuse must be made against the publishing of them; Since with respect to Reputation, I dare say, 'twill be no way to your disadvantage.

(Jones to Cotes, 25 October 1711 [4], Letter CV.)

I have received Mouton's Book and thank You for the Favour You did me in sending it. I have looked over what relates to his way of Interpolation, but I find no cause from thence to make any alteration. I beg Your pardon that I have not yet sent You my Second Paper. I had finished it before I received Your last letter with the Book, but since that time I have not had leasure to transcribe it. I design to send it with a paper concerning Logarithms which together with these other may serve to fill up a whole transaction, if You shall think them worthy to be published after you have read them.

(Cotes to Jones, 11 November 1711, Trinity College Library MS R 16 38, fo. 302.)

Towards the end of the year, Cotes had changed his mind, and had decided to publish his papers, other than Logometria, Part I, in a book to be printed at Cambridge. Jones sent Cotes a paper of Newton's on the construction of tables by a method of repeated sub-tabulation and, later, a folio manuscript by Nathanial Torperly and a quarto manuscript of Mercator's all relating 'in some measure, to the Method of Differences'. He thought it unlikely that these would contain anything of other than historical interest to Cotes, that neither Mercator, nor indeed anyone else, would have written anything 'so very considerable, as to deserve to accompany anything of yours'; and continued to press Cotes to let him have papers for publication in the *Philosophical Transactions of the Royal Society*. Cotes responded very courteously, finding Newton's paper 'well suited to those particular purposes for which he designed it', and did not doubt that he would find it very curious when he had 'leasure to examine it to the bottom'.

What I intend to print will make but a small Volume, I cannot say it will be bigger than that of Sr Isaac's which You lately published. It will contain the Lectures I have hitherto read in Publick, together with those which I shall read this Year, all of which amount to no more than Ten, for by the Statutes of my place I am obliged annually to make but two. I cannot indeed expect any profit from the Publication, twill be sufficient if the expenses of it can be defrayd. I have already put ye University to the charge of Types for some new characters which I have occasion to make use of & therefore for that reason as well as some others I cannot now draw back. What You mention that ye R: Society have chosen me one of their Members is altogether a peice of news to me. If it be so, I shall be very sensible of the Honour they have done me. That Title may recommend my papers to ye Publick though they be printed at Cambridge. If You insist upon my Promise of sending those things to You before they are printed I shall be ready to make it good. What I have further concerning the subject of differences consists of Ten Propositions whereof the Six first are particular & fitted for use and are sufficient for all cases that comonly happen, the other four are general. You will be able to judge of my Method by ye first proposition which I here send You. You may show it to Sr Isaac if You think it proper but I desire you would not show it to others.

(Cotes to Jones, undated but in reply to a letter from Jones dated 1 January 1712, accompanying the Newton paper on the construction of tables [4],
Letter CX.)

Canonotechnia is concerned with the systematic interpolation, i.e., sub-tabulation, and develops from Newton's Proposition III, in *Methodo Differentialis*. The method consists in deducing, from the entries in the central row of a difference table (equally-shaped pivotal values), the new entries which would result if interpolated values were present in the table. From these new differences, the interpolated values can be found. Cotes' method then, computes interpolated values from the entries in the difference table, just as modern interpolation methods do. As has been pointed out by D. T. Whiteside, formulae equivalent to both the Bessel and the Stirling interpolation formulae occur in Canonotechnia. (D. T. Whiteside, *The Mathematical Papers of Isaac Newton*, vol. 4 (8 vols., Cambridge, 1971), p. 61. I can only agree with Dr Whiteside that at least one of these formulae should be named after Cotes.

The procedure for sub-dividing a tabular interval into n equal parts is as follows. From the entries in the central row of a difference table, denoted by A, B, C, D, ..., a set of modified differences, A, B/N, C/n^2, D/n^3, ... is formed. These are denoted thus $⓪$, $①$, $②$, $③$, ... and called round differences. From the round differences, a further set of differences, called square differences, is formed. These are denoted thus; $\boxed{0}$, $\boxed{1}$, $\boxed{2}$, $\boxed{3}$, ... They are related to the round differences, and deduced from them, by a set of rules which it is the object of Canonotechnia to establish. These square differences are the new differences which would result if the required interpolated terms were present in the table. Therefore, from the square differences, the interpolates can be found. With Cotes' customary concern for the practical, the whole is reduced to a fairly simple procedure, and a chart provided to assist in the computation.

Cotes distinguishes between the two cases (a) an odd number, and (b) an even number of pivotal values, calling them medial and bimedial respectively (terms commonly used in connection with ratios). There are ten theorems. In the first six, Cotes considers bisection, trisection and quinquisection of the tabular interval, in both medial and bimedial cases. These are presented as practical rules and regarded as sufficient for most purposes. In the final four theorems, the results are generalised, and in a notable final scholium, Cotes discusses linear interpolation, with second-order and third-order corrections, and their limits of accuracy. Finally, in a postscript, the interpolates are expressed entirely in terms of round differences, in what later became known as Stirling's formula. It will be sufficient to give Proposition I in detail, and to illustrate it with a simple numerical example. Thus Proposition I is: *Given a series of equidistant terms of which one is midway between the others, let it be proposed to bisect the intervals by placing one term between each of the given terms, as accurately as possible.* Cotes uses Newton's notation thus, $4A$, $3A$, $2A$, $1A$, $A1$, $A2$, $A3$, $A4$, are the given terms. The values to be interpolated are $7a$, $5a$, $3a$, $1a$, $a1$, $a3$, $a5$, $a7$. The table shows the central three rows of the difference table, in which, as with Newton, differences are taken in the order $f_n - f_{n+1}$

$$
\begin{array}{llllllll}
4A \\
3A \\
2A \\
1A \\
\quad A & 1B & C & 1D & E & 1F & G & 1H \\
\quad\quad & B1 & & D1 & & F1 & & H1 \\
A1 \\
A2 \\
A3 \\
A4
\end{array}
$$

The centre row is completed as

$$A, B, C, D, E, F, G, H, \ldots,$$

where

$$B = \frac{1B + B1}{2}, \qquad D = \frac{1D + D1}{2}, \quad \text{and so on.}$$

Cotes now forms the round differences

$$①= A, \qquad ①=\frac{B}{2}, \qquad ②=\frac{C}{4}, \qquad ③=\frac{D}{8}, \quad \text{and so on.}$$

From these round differences, the square differences are formed according to the following scheme:

$$\boxed{10} = \boxed{\boxed{10}} - \cdots, \qquad\qquad \boxed{9} = \boxed{\boxed{9}} - \cdots,$$

$$\boxed{8} = \boxed{\boxed{8}} - \boxed{10} - \cdots, \qquad \boxed{7} = \boxed{\boxed{7}} - \tfrac{5}{4}\boxed{9} - \cdots,$$

$$\boxed{6} = \boxed{\boxed{6}} - \tfrac{3}{4}\boxed{8} - \tfrac{3}{10}\boxed{10} - \cdots, \qquad \boxed{5} = \boxed{\boxed{5}} - \boxed{7} - \tfrac{5}{16}\boxed{9}\cdots,$$

$$\boxed{4} = \boxed{\boxed{4}} - \tfrac{1}{2}\boxed{6} - \tfrac{1}{16}\boxed{8} - \cdots, \qquad \boxed{3} = \boxed{\boxed{3}} - \tfrac{3}{4}\boxed{5} - \tfrac{1}{8}\boxed{7}{}^{*},$$

$$\boxed{2} = \boxed{\boxed{2}} - \tfrac{1}{4}\boxed{4}{}^{*}, \qquad\qquad \boxed{1} = \boxed{\boxed{1}} - \tfrac{1}{2}\boxed{3}{}^{*},$$

$$\boxed{0} = \boxed{\boxed{0}}{}^{*},$$

(1)

* denoting termination of the series. As already mentioned, the square differences are the values which would replace $A, B, C, D, E, \ldots$, if the interpolates were present. The interpolated values are fairly simply obtained from the square differences, thus:

$$\frac{1a + a1}{2} = \boxed{0} + \tfrac{1}{2}\boxed{2},$$

$$\frac{1a - a1}{2} = \boxed{1},$$

$$\frac{3a + a3}{2} = \boxed{0} + 4 \cdot \tfrac{1}{2}\boxed{2} + 3 \cdot \boxed{4} + \tfrac{1}{2} \cdot \boxed{6},$$

$$\frac{3a - a3}{2} = 3 \cdot \boxed{1} + 4 \cdot \boxed{3} + \boxed{5},$$

(2)

and so on.

Consider as an illustrative example, the problem of bisecting the intervals in the table of integral cubes as under:

```
      1
      8
     27
             -37
     64             24
             -61            -6
    125             30            0
             -91            -6
    216             36
    343    -127
    512
    729
```

$A = 125,\ B = (-61 - 91)/2 = -76,\ C = 30,\ D = -6.$

$$\text{⓪} = A = 125,$$

$$\text{①} = \frac{B}{2} = -38,$$

$$\text{②} = \frac{C}{4} = \frac{15}{2},$$

$$\text{③} = \frac{D}{8} = -\tfrac{3}{4},$$

as above, and following Cotes' scheme, we have

$$\boxed{0} = \text{⓪} = 125,$$

$$\boxed{1} = \text{①} - \tfrac{1}{2}\boxed{3} = -37\tfrac{5}{8},$$

$$\boxed{2} = \text{②} = \tfrac{15}{2},$$

$$\boxed{3} = \text{③} = -\tfrac{3}{4}.$$

The interpolates can now be computed as follows:

$$\frac{1a + a1}{2} = \boxed{0} + \tfrac{1}{2}\boxed{2} = 128\tfrac{3}{4},$$

$$\frac{1a - a1}{2} = \boxed{1} \qquad = -37\tfrac{5}{8}.$$

From these two equations,

$$1a = \tfrac{729}{8} = (4\tfrac{1}{2})^3,$$

$$a1 = \tfrac{1123}{8} = (5\tfrac{1}{2})^3,$$

which are seen to be correct. Similarly, $a3$, $3a$; $a5$, $5a$ and so on can be computed.

In his round-difference and square-difference notation, Cotes has proposed a useful and practicable notation. The symbols are probably the special types he referred to in his letter to Jones, quoted above.

The next five propositions deal with interpolating 1, 2, 5 (and, by implication, 10) values between the pivotal values, in both medial and bimedial cases. Formulae such as (1) above for computing the square differences from the round differences are given in each case, as are formulae such as (2) for relating the square differences to the interpolates.

Before considering the origin of these, at first sight somewhat perplexing formulae, it will be helpful to consider Proposition II briefly.

Proposition II is: *To bisect the tabular intervals when the number of pivotal values is even*, i.e., the bimedial analogue of Proposition I. In this case, the three central rows of the difference table are as follows;

$$\begin{array}{ccccccccc} 1A & & 1C & & 1E & & 1G & \\ & B & & D & & F & & H \\ A1 & & C1 & & E1 & & G1 & \end{array}$$

and

$$A = \frac{1A + A1}{2}, \qquad C = \frac{1C + C1}{2}, \quad \text{and so on.}$$

The round differences, as before, are

$$\textcircled{0} = A, \qquad \textcircled{1} = \frac{B}{2}, \qquad \textcircled{2} = \frac{C}{4}, \dots$$

The square differences are derived from these according to the scheme shown in (3) below and the interpolates computed from the relations (4).

$$\boxed{10} = \textcircled{10} - \cdots, \qquad\qquad \boxed{9} = \textcircled{9} - \cdots,$$

$$\boxed{8} = \textcircled{8} - \tfrac{3}{2}\boxed{10} - \cdots, \qquad \boxed{7} = \textcircled{7} - \tfrac{3}{4}\boxed{9} - \cdots,$$

$$\boxed{6} = \textcircled{6} - \tfrac{5}{4}\boxed{8} - \tfrac{9}{16}\boxed{10} - \cdots, \qquad \boxed{5} = \textcircled{5} - \tfrac{1}{2}\boxed{7} - \tfrac{1}{16}\boxed{9}^*,$$

$$\boxed{4} = \textcircled{4} - \boxed{6} - \tfrac{5}{16}\boxed{8} - \tfrac{1}{32}\boxed{10}^*, \qquad \boxed{3} = \textcircled{3} - \tfrac{1}{4}\boxed{5}^*, \qquad\qquad (3)$$

$$\boxed{2} = \textcircled{2} - \tfrac{3}{4}\boxed{4} - \tfrac{1}{6}\boxed{6}^*, \qquad \boxed{1} = \textcircled{1}^*,$$

$$\boxed{0} = \textcircled{0} - \tfrac{1}{2}\boxed{2}^*,$$

$$a = \boxed{0}^*,$$

$$\frac{2a + a2}{2} = \boxed{0} + 2\boxed{2} + \tfrac{1}{2}\boxed{4}^*,$$

$$\frac{2a - a2}{2} = 2\boxed{1} + \boxed{3}^*,$$

$$\frac{4a + a4}{2} = \boxed{0} + 8\boxed{2} + 10\boxed{4} + 4\boxed{6} + \tfrac{1}{2}\boxed{8}^*,$$

$$\qquad\qquad\qquad\qquad\qquad\qquad\qquad\qquad (4)$$

$$\frac{4a - a4}{2} = 4\boxed{1} + 10\boxed{3} + 6\boxed{5} + \boxed{7}^*,$$

$$\frac{6a + a6}{2} = \boxed{0} + 18\boxed{2} + 52\tfrac{1}{2}\boxed{4} + 56\boxed{6} + 27\boxed{8} + 6\boxed{10} + \cdots,$$

$$\frac{6a - a6}{2} = 6\boxed{1} + 35\boxed{3} + 56\boxed{5} + 36\boxed{7} + 10\boxed{9} + \cdots,$$

and the series is completed as

$$5a, 4a, 3a, 2a, 1a, a, a\,1, a\,2, a\,3, a\,4, a\,5.$$

Similar results are given for the cases considered in the Propositions III to VI. Cotes observes, in a note at the end of Proposition VI, that the results so far given are usually sufficient for the construction of tables; and proceeds to the general case.

In the course of this generalisation, Cotes gives a useful algorithm for computing the coefficients in formulae such as (1) and (3) above. The square differences being relatively simply obtained in this way, their relationship to the required interpolates is shown in two tables (Fig. 50). Cotes' instructions for the use of these tables are: 'For all medial cases, and for the bimedial case, n even [n being the number of interpolates] use the Tabula Posterior for the semi-sum, and the tabula posterior for the semi-difference; reverse this for the bimedial case, n odd.'

In the Scholium to Proposition IV (followed by a postscript) Cotes, for the first time, gives some indication of his methods. Here he reverts to the general problem of interpolation, adding that correction by second, third and higher-order differences is usually possible, but not always useful. He will, he says, add 'a not inelegant method', which can be derived from Newton's interpolation formula given in Proposition III of *De Methodo Differentiali*.

The centre three rows of a bimedial table are given, and the value Z is to be interpolated between P and N, distant l from L, m from M, n from N, and so on.

R						
Q		1C		1E		
P	B		D		F	
N		C1		E1		
M						
L						

Cotes simply states

$$Z = N + nB - \frac{n}{2}C - \frac{nop}{6}D + \frac{mnpq}{24}E + \frac{mnopq}{120}F - \frac{lmnpqr}{720}G -, \ldots$$

(note, $o = n - (1/2) = -m/2$). In this, putting $n + p = 1$, $m = 1 + n$, $1 = 2 + n$, and so on, and $Z = f(x)$, $N = f_0$, $P = f_1$, etc., and allowing for the

TABULA PRIOR

Class	Term	0	1	2	3	4	5	6	7	8	9	10
I	Semisum Bimed. $\frac{1a+a1}{2}$	1										
I	Semidiff. Medial $\frac{1a-a1}{2}$		1									
II	Semisum Bimed. $\frac{2a+a2}{2}$	1		1								
II	Semidiff. Medial $\frac{2a-a2}{2}$		2	1								
III	Semisum Bimed. $\frac{3a+a3}{2}$	1		3		1						
III	Semidiff. Medial $\frac{3a-a3}{2}$		3	4	1							
IV	Semisum Bimed. $\frac{4a+a4}{2}$	1		6		5		1				
IV	Semidiff. Medial $\frac{4a-a4}{2}$		4	10	6	1						
V	Semisum Bimed. $\frac{5a+a5}{2}$	1		10		15		7		1		
V	Semidiff. Medial $\frac{5a-a5}{2}$		5	20	21	8	1					
VI	Semisum Bimed. $\frac{6a+a6}{2}$	1		15		35		28		9		1
VI	Semidiff. Medial $\frac{6a-a6}{2}$		6	35	56	36	10					
VII	Semisum Bimed. $\frac{7a+a7}{2}$	1		21		70		84		45		11
VII	Semidiff. Medial $\frac{7a-a7}{2}$		7	56	126	120	55					
VIII	Semisum Bimed. $\frac{8a+a8}{2}$	1		28		126		210		165		66
VIII	Semidiff. Medial $\frac{8a-a8}{2}$		8	84	252	330	220					
IX	Semisum Bimed. $\frac{9a+a9}{2}$	1		36		210		462		495		266
IX	Semidiff. Medial $\frac{9a-a9}{2}$		9	120	462	792	715					
X	Semisum Bimed. $\frac{10a+a10}{2}$	1		45		330		924		1287		1001
X	Semidiff. Medial $\frac{10a-a10}{2}$		10	165	792	1716	2002					

TABULA POSTERIOR

Class	Term	0	1	2	3	4	5	6	7	8	9	10
I	Semidiff. Bimed. $\frac{1a-a1}{2}$		½									
I	Semisum Medial $\frac{1a+a1}{2}$	1		½								
II	Semidiff. Bimed. $\frac{2a-a2}{2}$		1½		½							
II	Semisum Medial $\frac{2a+a2}{2}$	1		2		½						
III	Semidiff. Bimed. $\frac{3a-a3}{2}$		2½		2½		½					
III	Semisum Medial $\frac{3a+a3}{2}$	1		4½		3		½				
IV	Semidiff. Bimed. $\frac{4a-a4}{2}$		3½		7		3½		½			
IV	Semisum Medial $\frac{4a+a4}{2}$	1		8		10		4		½		
V	Semidiff. Bimed. $\frac{5a-a5}{2}$		4½		15		13½		4½		½	
V	Semisum Medial $\frac{5a+a5}{2}$	1		12½		25		17½		5		½
VI	Semidiff. Bimed. $\frac{6a-a6}{2}$		5½		27½		36½		22		5½	
VI	Semisum Medial $\frac{6a+a6}{2}$	1		18		52½		56		27		6
VII	Semidiff. Bimed. $\frac{7a-a7}{2}$		6½		45½		91		76		32½	
VII	Semisum Medial $\frac{7a+a7}{2}$	1		24½		96		147		105		36½
VIII	Semidiff. Bimed. $\frac{8a-a8}{2}$		7½		70		169		225		137½	
VIII	Semisum Medial $\frac{8a+a8}{2}$	1		32		166		336		330		176
IX	Semidiff. Bimed. $\frac{9a-a9}{2}$		8½		102		357		561		467½	
IX	Semisum Medial $\frac{9a+a9}{2}$	1		40½		270		693		891		643½
X	Semidiff. Bimed. $\frac{10a-a10}{2}$		9½		142½		627		1254		1358½	
X	Semisum Medial $\frac{10a+a10}{2}$	1		50		412½		1320		2145		2002

Fig. 50. Table showing the relationship between Cotes' square differences, and interpolated values. From *Harmonia Mensurarum*, Cambridge University Press, 1722.

fact that Cotes, as with Newton, takes his differences in the order $f_n - f_{n+1}$, this interpolation formula reduces to (using modern central difference notation):

$$f(x) = f_0 + \theta \delta f_{\frac{1}{2}} + \frac{\theta(\theta-1)}{2!}\left(\frac{\delta^2 f_0 + \delta^2 f_1}{2}\right)$$
$$+ \frac{\theta(\theta-\frac{1}{2})(\theta-1)}{3!}\delta^3 f_{\frac{1}{2}} + \cdots, \tag{5}$$

i.e., the formula now known as the Bessel or Newton–Bessel interpolation formula (F. W. Bessel 1784–1846). In this formula I have replaced n with θ to avoid confusion with n the number of interpolates in a tabular interval. The formula is symmetrical about the mid-point of the tabular interval, thus giving equal weight to the entries on either side of that point: it is the source of Cotes' formulae for the square differences in the bimedial cases, as will be shown below. Cotes comments that, in using the formula for linear interpolation, errors will not normally exceed one eighth of the second difference and $\sqrt{3}/216$ of the second difference. These are the maximum values of the relevant coefficients in the formula, namely, $[\theta(\theta-1)]/2$, and $[\theta(\theta-\frac{1}{2})(\theta-1)]/3!$.

It will not have escaped the notice of the perceptive reader, that, since the round differences determine the square differences, and the square differences determine the interpolates, then it should be possible to express the interpolates solely in terms of round differences. Cotes takes up this point in a postscript, giving

$$\frac{1a+a1}{2} = ⓪ + \tfrac{1}{2}② - \tfrac{1}{8}④ + \tfrac{1}{16}⑥ - \tfrac{5}{128}⑧, \quad \text{and so on,} \tag{6}$$

for Proposition I, and a similar result for the semi-difference, with corresponding results for all the Propositions I–VI. These results, says Cotes, are obtained from Newton's Propositions by a suitable rearrangement. Cotes gives the result, but not the rearrangement, which is to restate the Proposition III in a form symmetrical about the mid-tabular entry. With this change, Newton's interpolation formula becomes, in modern notation,

$$f(x) = f(0) + \theta(\delta f_{-\frac{1}{2}} + \delta f_{\frac{1}{2}}) + \frac{\theta^2}{2}\delta^2 f_0$$

$$+ \frac{\theta(\theta^2-1)}{3!}\frac{(\delta^3 f_{-\frac{1}{2}} + \delta^3 f_{\frac{1}{2}})}{2} + \frac{\theta^2(\theta^2-1)}{4!}\delta^4 f_0$$

$$+ \frac{\theta(\theta^2-1)(\theta^2-2)}{5!}\frac{(\delta^5 f_{-\frac{1}{2}} + \delta^5 f_{\frac{1}{2}})}{2} + \cdots, \tag{7}$$

which is the Newton–Stirling interpolation formula, given by James Stirling (*c.* 1696–1770) in the *Philosophical Transactions of the Royal Society*, 30 (1719), 1050. Thus Stirling has priority of publication, but Cotes is the prior user. Putting $\theta = \frac{1}{2}$ and $\theta = -\frac{1}{2}$ in turn in (7), and averaging the two results, we have (6), and similarly by putting $\theta = 1\frac{1}{2}$ and $\theta = -1\frac{1}{2}$ in turn and averaging, and so on, we have the other formulae for calculating the interpolates in terms of round differences, all of which Cotes gives. A similar procedure applied to (5), the 'Bessel' formula, gives the interpolates in terms of round differences for the bimedial cases.

Finally then, we come to the origin of the square-difference formulae. If we seek to follow Cotes in developing these, we can proceed as follows: Consider a medial table, central tabular entry f_0, and the n interpolates between f_0 and f_1 denoted by $f_{1/n}$, $f_{2/n}$, etc., and those between f_0 and f_{-1} by $f_{-1/n}$, $f_{-2/n}$, etc. Then

$$\boxed{0} = f_0,$$

$$\boxed{1} = \frac{f_{1/n} - f_{-1/n}}{2},$$

$$\boxed{2} = f_{1/n} - 2f_0 + f_{-1/n},$$

$$\boxed{3} = \frac{f_{2/n} - f_{-2/n}}{2} - (f_{1/n} - f_{-1/n}),$$

$$\boxed{4} = f_{2/n} - 4f_{-1/n} + 6f_0 - 4f_{1/n} + f_{-2/n}$$
$$= f_{2/n} + f_{-2/n} - 4(f_{1/n} + f_{-1/n}) + 6f_0.$$

Also, substituting $\theta = 1/n$, $-1/n$, $2/n$, $-2/n$ in turn in (7), we obtain four more equations, thus

$$\boxed{1} = \text{①} + \frac{1 - n^2}{3!}\text{③} + \frac{(1 - n^2)(1 - 4n^2)}{5!}\text{⑤} + \cdots$$

$$\boxed{2} = \text{②} + \frac{2(1 - n^2)}{4!}\text{④} + \frac{2(1 - n^2)(1 + n^2)}{6!}\text{⑥} + \cdots$$

$$\boxed{3} = \text{③} + \frac{30(1 - n^2)}{5!}\text{⑤} \cdots,$$

$$\boxed{4} = \text{④} + \frac{(1 - n^2)}{6!}\text{⑥} \cdots.$$

Eliminating the round differences as required between these sets of equations, we arrive at

$$\boxed{1} = \text{①} - \frac{n^2 - 1}{3!}\boxed{3} - \frac{(n^2 - 1)(n^2 - 4)}{5!}\boxed{5},$$

$$\boxed{3} = \text{③} - \frac{5n^2 - 5}{4 \cdot 5}\boxed{5},$$

$$\boxed{2} = \text{②} - \frac{n^2 - 1}{3 \cdot 4}\boxed{4} - \frac{(n^2 - 1)(n^2 - 4)}{3 \cdot 4 \cdot 5 \cdot 6}\boxed{6},$$

$$\boxed{4} = \textcircled{4} - \frac{5n^2 - 5}{5 \cdot 6}\boxed{6}.$$

These agree with the entries in Cotes' generalised results, which he states without proof in his Proposition VII. The general results for the bimedial cases are similarly derived, using the interpolation formula which is symmetrical about the mid-point of the tabular interval, i.e., (5) – the Bessel formula. Thus, to revert to Proposition I, which is seen to be a medial case, with $n = 2$, we have from the above general formulae,

$$\boxed{1} = \boxed{1} - \tfrac{1}{2}\boxed{3},$$

$$\boxed{2} = \textcircled{2} - \tfrac{1}{4}\boxed{4},$$

$$\boxed{3} = \textcircled{3} - \tfrac{3}{4}\boxed{5},$$

$$\boxed{4} = \textcircled{4} - \tfrac{1}{2}\boxed{6}.$$

Cotes continues the sequences somewhat further.

Thus, to conclude, although interpolation is readily carried out by the Bessel or Stirling formulae, for purposes of sub-tabulation, Cotes' square-difference formulae avoid the need for repeated application of the interpolation formulae. Once the square differences have been computed, the interpolates can be quickly written in with the aid of the Tabula Prior and the Tabula Posterior. The whole constitutes a sub-tabulation procedure of considerable utility, particularly if all computations have to be done by hand. As with his work on integrals, Cotes is just ahead of the field. His notation is ingenious, his descriptions clear, his proofs sparse or absent. Delay in publication meant that Brook Taylor (who was active in urging Robert Smith to get on and publish Cotes' work) published his *Methodus Incrementorum Directa et Inversa* (London, 1715) first. Stirling's *Differential Method* followed in 1730. Neither Taylor nor Stirling referred to Cotes' work. In her excellent article on Cotes in the *Dictionary of National Biography*, Agnes Mary Clarke states that Canonotechnia was translated into French by La Caille in 1741 and appeared in *Histoire de l'Académie Royale des Sciences* (Paris), p. 238. Here she is, unusually, mistaken; the work concerned was Aestimatio Errorum. In 1778 Lagrange drew attention to Cotes' work in the *Histoire et Mémoires de l'Académie Royale de Berlin* (Berlin), p. 111–61, otherwise this very substantial achievement seems to have aroused little public attention.

7

The missing works

Referring again to Smith's Preface to *Harmonia Mensurarum* (Cambridge, 1722), we find that he listed other works by Cotes, not yet published. He cited 'A Compendium of Arithmetic', 'A Treatise on the Solution of Equations', 'A General Dioptric Theorem' and a tract 'On the Nature of Curves', 'all of which I intend to Publish'.

The 'Compendium of Arithmetic' seems to be lost without trace. Among the Clare papers are two which probably formed part, or possibly the whole, of the work on the Solution of Equations. One is headed 'Aequationis Biquadraticae ope Cubicae Resolutio' and sets out the general method for doing this; the other, headed 'Resolutio Aequationem Affectorum in Numeri, Auth. D^{mo} R. Cotes', is a statement and proof of the method of solution of polynomial equations by the Newton–Raphson method. Cotes adds a note on reducing the labour of computation by not carrying more decimal places at each stage of the computation than the problem requires. These papers are folded up with sheets of commentary on Newton's *Algebra* (later published as *Arithmetica Universalis*, ed. W. Whiston (Cambridge, 1707)). Cotes at one time hoped to encourage Newton to publish a revised edition of his *Algebra*, and the commentaries were probably prepared with this in view. They are concerned with the 1707 Latin edition (Cotes' own manuscript copy of Newton's algebra lectures, which formed the substance of the book, is in the Trinity College Library MS R 16 39). The most extensive commentaries are on the sections 'On the Finding of Divisors' (i.e., the factorisation of polynomials), and 'On the Extraction of Roots' (square and cube roots of surd quantities). In 1727 G. J. 's Gravesande published *Matheseos Universalis Elementa* (Leyden, 1727), which includes on pages 175–230, commentary on precisely the same sections, following much the same lines as Cotes' commentary. Colin Maclaurin, in the Preface to his *Treatise of Algebra* (London, 1748) says that Gravesande's intention

Fig. 51. *S* is a rational curve. *Pm* is the harmonic mean of *PA*, *PB*, *Pc*, etc.

was to illustrate the kind of commentary needed on Newton's *Algebra*, and adds that his own *Treatise* is designed to serve as just such a commentary. Once again, Cotes is ahead of the field, but failed to publish. What is probably part of Cotes' 'Tract on the Nature of Curves' appears in Part III of Maclaurin's *Treatise*, in the Appendix. It is the famous 'Theorem on harmonic means'. In Maclaurin's own words:

If any right line drawn from a point *P* cut a geometrical line in as many points as it has dimensions, in which let *Pm* be always taken an harmonical mean between all the segments of the drawn line, terminated by the point *P* and the curve, the point *m* will be in a right line, and this is Cotes' theorem, or nearly related to it.

That is, if a line *s* through a fixed point *P* has its full quota of intersections *A*, *B*, *C*, *D*, ..., with a rational curve *S*, and P_m is the harmonic mean of *PA*, *PB*, *PC*, ..., then the locus of *m* as *s* varies in a pencil through *P*, is a straight line (Fig. 51). Maclaurin proves the theorem (he acknowledges the theorem but not the proof from Cotes) at great length, making use of elementary symmetric functions of the roots of polynomials, the harmonic properties of the complete quadrilateral, and some results from his own *Treatise of Fluxions* (2 vols., Edinburgh, 1742).

The Theorem on harmonic means was said by W. W. Rouse Ball (*A History of the Study of Mathematics at Cambridge* (Cambridge, 1889), p. 90) to have given rise to the title *Harmonia Mensurarum*: a surprising error from a scholar of such distinction.

A study of Maclaurin's *Algebra*, and the brief description of Cotes' tract on curves given by Smith, reveals a number of similarities. Smith says that Cotes dealt algebraically with conics, and showed that many of their properties were special cases of properties of higher curves. He also added (says Smith) constructions of loci and equations, and the solution of many problems dependent on them. In the Preface to Maclaurin's *Algebra*, probably written by Martin Folkes, we find

The Latin appendix is a proper sequel and high improvement of what had been demonstrated in part III concerning the relation of curve lines and equations, a subject with which our author, [i.e., Maclaurin] had been early acquainted, witness his *Geometrica Organica* [1719] which gained him great

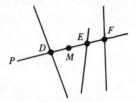

Fig. 52. *PM* is the harmonic mean of *PD, PE, PF.*

distinction. Yet he frankly owns he was led to many of the propositions in this appendix, from a theorem of Mr Cotes, communicated to him without any demonstration by the Reverend and learned Dr Smith, Master of Trinity College, Cambridge. How he has profited from that, the learned will judge. He himself set some value on this performance.

The idea that properties of conics are special cases of properties of higher curves receives quite a lot of attention from Maclaurin, and in Article 56 the following result, concerning three straight lines appears (Fig. 52). A straight line through a fixed point *P* cuts three straight lines in *D, E, F,* and *PM* is the harmonic mean of *PD, PE, PF*: 'Then the locus of *M* will be a right line, and this is a property of three lines invented by Cotes.' Poncelet gave a proof of the general theorem, founded on a continuity argument (see M. Marie, *Histoires des Sciences Mathématique et Physique*, vol. 7 (12 vols., Paris, 1885), p. 222). Other writers have pointed out that it is not necessary for the line through *P* to have its full quota of intersections with a rational curve, for the theorem to remain true (a consequence of the fact that complex roots of polynomial equations occur in conjugate pairs).

Finally, to Cotes' general Dioptic Theorem. There is a manuscript copy of this among the Clare papers, and Smith published it, with due acknowledgement to Cotes, in his *Compleat System of Optics in Four Books* vol. I (2 vols., Cambridge, 1738), p. 76–8. The Theorem is: *To find the apparent magnitude, situation, apparent place and degree of distinctness, with which an object is seen through any number of lenses, of any sort, at any distance from each other, and from the eye and object.* Smith comments:

That noble and beautiful theorem, from which I deduced all these corollaries, was the last invention of that great Mathematician, Mr Cotes, just before his death at the age of 32, upon which occasion I am told that Sir Isaac Newton said, if Mr Cotes had lived, we might have known something. His demonstration of it is likewise so elegant and clear that it highly deserved to follow the theorem but ... I was obliged to substitute another in its stead, and to give Mr Cotes' by itself.

This comment by Smith is the source of the attribution to Newton of the often quoted remark: 'If Mr Cotes had lived, we might have known something.' I am indebted to Dr Whiteside for pointing out to me that Smith wrote this comment on the title page of his own copy of the *Harmonia Measurarum* (Trinity College Library MS Adv b 1 15).

Cotes' (and Smith's) proofs are readily available, and it is not necessary to reproduce them here. Lagrange, writing in 1778, acknowledged Cotes as being one of the two geometers who had attempted a general theory of lenses (the other was Euler). See J. L. Lagrange, 'Sur La Theorie des Lunettes', *Histoire et Memoires, Academie Royale de Berlin* (Berlin, 1778), p. 162.

8

Conclusion

Almost all contemporary comment on Cotes' personality refers to his amiability, courtesy, and sweetness of temper; allowing for eighteenth-century conventions of expression, these seem indeed to have been his outstanding traits. Industrious, dutiful, intensely loyal to the established order of things, he seems one of the least revolutionary apostles of the Age of Enlightenment. Yet his strong and undeviating admiration and respect for Newton gave his anti-Cartesianism a partisan flavour. In Bentley's controversies, Cotes (Bentley's protégé) consistently supported Bentley, the established Master (support which probably cost him a lectureship, see [4], p. 203). In religion he remained within the established church, receiving Holy Orders in 1713.

Cotes was no dreamy academic, but a competent man of affairs, with a good practical sense. He served his College as junior bursar from 1707 to 1710, and presided, for example, over the day-to-day matters concerning the construction, equipping and bringing into operation, of the observatory, managing the fund-raising for this project. He seems to have carried out his duties efficiently, although at the time of his unexpected death he owed the fund some money: the record book which he kept is not extant. Nevertheless, under his guidance the observatory was established, and although not completed, was brought into operation. The amount of practical astronomy recorded as carried out there is disappointingly small. Under Robert Smith, Cotes' successor, the money came in more slowly and the building was not completed until 1739. The observatory seems, after Cotes, to have been used mainly as a mechanics lecture room, Smith continuing the course of mechanics which Cotes and Whiston had established. Science and Mathematics were certainly not neglected in the lectures delivered at Oxford and Cambridge during the seventeenth and eighteenth centuries. (See, for example, R. Gunter, *Early*

Science at Oxford (14 vols., Oxford, 1923–1945); also, *Early Science at Cambridge* (1 vol., Oxford, 1937).) Cotes and Whiston, in their course of hydrostatical and pneumatical lectures, mostly written by Cotes, were among the earliest Newtonians to teach by means of practical experiments which their hearers were obliged to carry out, usually before hearing the lecture. Robert Smith published the lectures, (Cotes R. *Hydrostatical and Pneumatical Lectures*, first edition (Cambridge, 1738)) and wrote in his Preface,

the method, in general, of teaching Philosophy by Courses of Experiments . . . is now so much practised and approved of by the most eminent Professors all over Europe [and], has so greatly contributed to the propogation and increase of knowledge, in the little time it has been duly cultivated, that nothing more need be said to show the usefulness and excellence of it.

The sixteen lectures were delivered as a four-week course, and arranged under four headings which briefly indicate their scope:

(1) Hydrostatical tryals and conclusions.
(2) Pneumaticks illustrated by experiments for the most part tubular, being such as were wont to be made before the air pump was invented.
(3) Most known properties of the air established by the air pump, and other engines.
(4) The more hidden properties of the air considered by the help of the like engines.

In his written Introduction to the lectures, Cotes explained carefully why he thought this the best teaching order. Smith refers further to 'the general satisfaction they have given, and the great and established reputation of their author'. The lectures are written in a simple and straightforward style, the mathematical requirements are kept to a minimum, and reference is made to historical developments and to current works and discoveries. Altogether, they mark a notable advance in the teaching of science at undergraduate level.

By 1792, according to Willis & Clarke (1886), the Plumian trustees found that 'The Professor had neither occupied the said rooms and leads, nor fulfilled the conditions for at least 50 years.' And that 'the Observatory, and the instruments belonging to it, was, through disuse, neglect and want of repair, so much dilapidated as to be entirely unfit for the purpose intended', and they relinquished their claim to it. It was taken down in 1797. Lord Trevelyan, giving the Master's speech at the Commemoration of Benefactors dinner in 1942, said: 'When the young man became the first Plumian Professor of Astronomy, Bentley erected for him the Observatory on the top of the Great Gate.

Its cupola disfigured the beauty of the Gate until its removal in 1797, but it was the only Observatory in Cambridge.'

In another project of Bentley's, the refurbishing and decorating of the College chapel, Cotes was again generally responsible for the practical supervision. A story exists of Bentley, feeling it necessary at last to show himself in chapel, being unable to enter his pew because, from long disuse, the lock had rusted; a locksmith had to be fetched.

Industrious, dutiful and loyal, Cotes could be firm and decisive, and even at times a little acerbic. Firmness and patience characterise his correspondence with Newton. Clarity and decision show in his advice, for example to Whiston, as in the following letter where Cotes shows his good practical sense. Whiston had put forward a scheme for a survey of the country, based on observations of flares' discharged from mortars, to a height of a mile or so. In the early trials the mortars were on Hampstead Heath. Whiston claimed Halley's support for the scheme; Cotes had his doubts.

I have hitherto sent you no account of your light, because I could not yet see it. I wish You good success in Your Enterprise. I fear the distance of 60 miles is too great for Your mortar. It will be very difficult for You to find Persons in every Market Town capable of making the requisite Observations. If You intend to leave the execution of Your project to such Persons as You think the least unfit in every town: You should, before You begin print a small book & teach 'em exactly & clearly how to draw a Meridian, how to take the bearing of Your Light from that Meridian, & how to observe the interval of the Sound. Unless they be precisely and punctually instructed in these things, and be furnished with such instruments as You shall think proper, and moreover, be faithful and diligent in the work: Your map must needs be but a lame Performance. I am glad to hear from You that Dr Halley puts the thing forward. I wish you would follow his advice intirely, who is certainly best able to advise You how to proceed.

Your

Humble Servant

R Cotes.

I would advise You to leave out that Paragraph about the Measure of the Earth. When You speak of doing this more exactly than the French, I am apt to think You have not sufficiently composed the thing.

(Cotes to Whiston, 2 December 1714 [4], Letter XXV.)

But it was for his mathematical work that Cotes used his best creative energies. Cautious and uncertain at first about the value of his own work, he came to have sufficient confidence in its worth to resist William Jones' urging to publish in the *Philosophical Transactions of the Royal Society* and to plan a major publication at Cambridge, at the University Press which Bentley had done so much to restore. As has been shown in earlier chapters, he made substantial advances in the theory of logarithms, in the integral calculus and coordinate geometry, and in numerical methods, particularly interpolation and table construction.

On Cotes' early death, on 5 June 1716, Bentley composed the epitaph which is inscribed on the memorial in Trinity College Chapel. The epitaph was reproduced in *Harmonia Mensurarum*, and in his own copy Robert Smith wrote: 'Sir Isaac Newton, speaking of Mr. Cotes, said "if he had lived we might have known something".'

Had Cotes lived, what might we have known? Cotes would certainly have followed publication of his book with papers in the *Philosophical Transactions of the Royal Society* and further publication of some, at least, of his minor works. He had the position and the ability to become the centre of a widening circle of friends and correspondents which would have included at least De Moivre, Taylor and, in due course, the rising young Stirling and Maclaurin. His correspondence would probably have extended to the continent of Europe. Although his Preface to *Principia* had not helped relations, Cotes gained some repute in France, both Aestimatio Errorum and the *Hydrostatical and Pneumatical Lectures* (second edition, Cambridge, 1747) being translated into French. It is unlikely that Brook Taylor would have indulged in his irritating 'integral challenge', instead we might have hoped for a more fruitful correspondence on methods of integration. We could reasonably have expected Cotes to make further contributions to complex numbers, to clearer recognition of circular and hyperbolic functions and their inverses, and possibly readier recognition of the power of the Continental methods.

It is tempting, and reasonable, to speculate that the post-Newtonian hiatus in the development of British mathematics would have been at least delayed in its onset if not altogether avoided. As it was, not only was there no one to succeed Newton, there was no one to succeed Cotes. His death struck a blow at Bentley's plans; the epitaph reflects a genuine sorrow.

Let Bentley, Smith and Newton have the last words (Fig. 53):

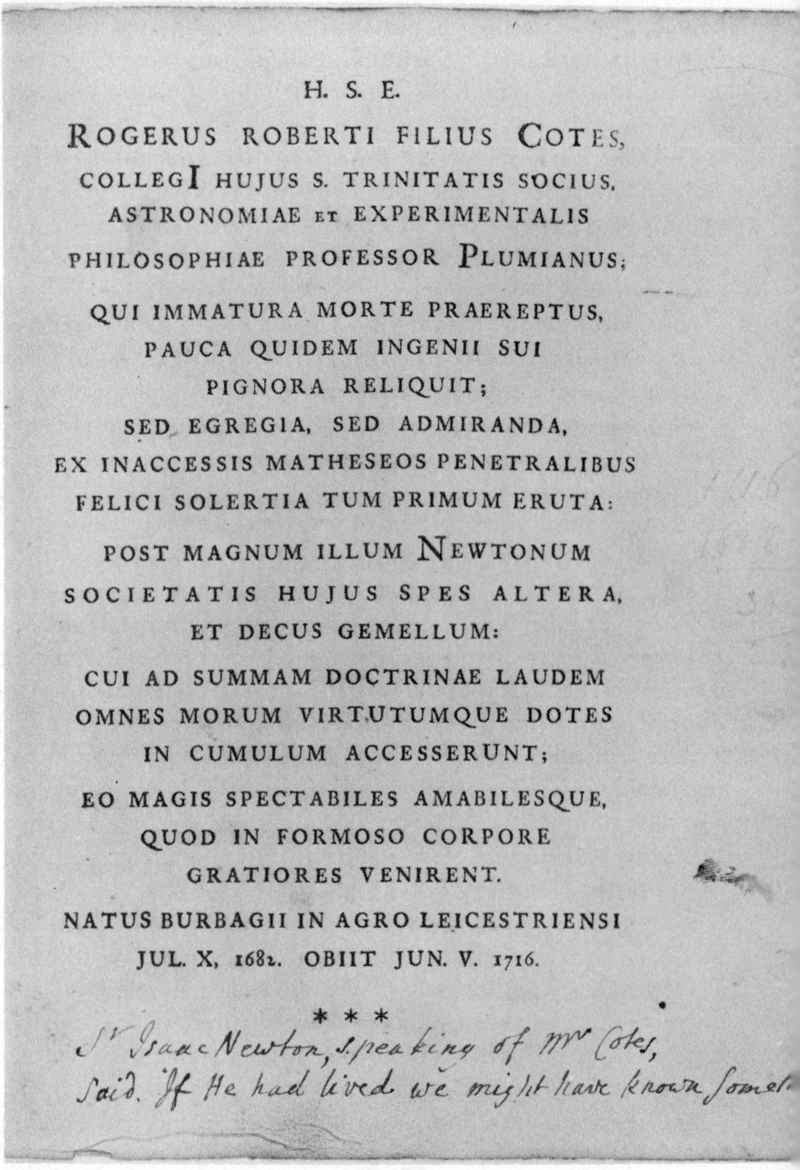

H. S. E.

ROGERUS ROBERTI FILIUS COTES,
COLLEGI HUJUS S. TRINITATIS SOCIUS,
ASTRONOMIAE ET EXPERIMENTALIS
PHILOSOPHIAE PROFESSOR PLUMIANUS;

QUI IMMATURA MORTE PRAEREPTUS,
PAUCA QUIDEM INGENII SUI
PIGNORA RELIQUIT;
SED EGREGIA, SED ADMIRANDA,
EX INACCESSIS MATHESEOS PENETRALIBUS
FELICI SOLERTIA TUM PRIMUM ERUTA:

POST MAGNUM ILLUM NEWTONUM
SOCIETATIS HUJUS SPES ALTERA,
ET DECUS GEMELLUM:

CUI AD SUMMAM DOCTRINAE LAUDEM
OMNES MORUM VIRTUTUMQUE DOTES
IN CUMULUM ACCESSERUNT;

EO MAGIS SPECTABILES AMABILESQUE,
QUOD IN FORMOSO CORPORE
GRATIORES VENIRENT.

NATUS BURBAGII IN AGRO LEICESTRIENSI
JUL. X, 1682. OBIIT JUN. V. 1716.

* * *

*Sr Isaac Newton, speaking of Mr Cotes,
said. If He had lived we might have known somet*

Fig. 53. (By permission of Trinity College Library, Cambridge.)

APPENDIX 1

English translation of Logometria

PROPOSITION I

To find the measure of any ratio whatever

Let a ratio be proposed between AC and AB, the measure of which must be determined. Consider the distance between the boundaries B, C [Fig. 1] as divided into innumerable small particles of the smallest possible size such as PQ, and the ratio between AC and AB as divided into the same number of ratios of the smallest possible size such as that between AQ and AP: and if the magnitude of the ratio between AQ and AP is given; division will give the ratio of PQ to AP; and also the given magnitude of the ratio between AQ and AP can be shown from the given quantity $\frac{PQ}{AP}$. AP remaining unchanged, consider the particle PQ to grow or to diminish in any proportion whatever, and the ratio between AQ and AP will grow or diminish in the same proportion: if the particle is taken as doubled or trebled, halved or divided by three, the ratio will also become doubled or trebled, halved or divided by three; it will still be shown therefore by the quantity $\frac{PQ}{AP}$. But it can also be shown by $M\frac{PQ}{AP}$ where any fixed quantity M is assumed; the quantity $M\frac{PQ}{AP}$ will therefore be a measure of the ratio between AQ and AP. This measure will indeed have a different magnitude, and will be accommodated to a different system, according to the different magnitude assumed for the quantity M, which for this reason should be called the Modulus of the System. Now just as the sum of all the ratios between AQ and AP is equal to the proposed ratio, which AC certainly has to AB, so the sum of all the measures (which can be discovered by fairly well known methods) will be equal to the required measure of that same proposed ratio.

143

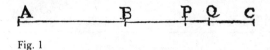

Fig. 1

Corollary 1
Take AP and AQ so nearly equal that their difference PQ is as small as possible; $M\dfrac{PQ}{AP}$ or $M\dfrac{PQ}{AQ}$ will be equal to the measure of the ratio between AQ and AP to modulus M.

Corollary 2
Whence the modulus M is to the measure of the ratio between AQ and AP as the length of either AP or AQ is to the difference PQ.

Corollary 3
Given the ratio between AC and AB, the sum of all $\dfrac{PQ}{AP}$ is given, and the sum of all $M\dfrac{PQ}{AP}$ is as M. Hence the measure of any given ratio whatever is as the chosen modulus of the system.

Corollary 4
The modulus therefore in all systems of measure, is always equal to the measure of a fixed and immutable ratio. So that I accordingly call it the *Ratio Modularis* [Modular Ratio].

Scholium I
It will be made clear by the solution of a problem as an example. Let z be some quantity, determined and constant, and let x be an undetermined and continuously varying quantity, whose fluxion is $\dot{x}$; and the ratio between $z+x$ and $z-x$ is required. Set this ratio equal to the ratio between y and 1, and further, let the number y be represented by AP, its fluxion $\dot{y}$ by PQ, 1 by AB; and from corollary 1 it will be deduced that the fluxion of the required measure of the ratio between y and 1 is $M\dfrac{\dot{y}}{y}$. Replace now for y its value $\dfrac{z+x}{z-x}$, and also for $\dot{y}$ the fluxion of the value $\dfrac{2z\dot{x}}{(z-x)^2}$ and the fluxion of the measure will emerge as $2M\times\dfrac{z\dot{x}}{zz-xx}$ or $2M\times\dfrac{\dot{x}}{z-\dfrac{xx}{z}}$ or $2M$ times $\dfrac{\dot{x}}{z}+\dfrac{\dot{x}x^2}{z^3}+\dfrac{\dot{x}x^4}{z^5}+$etc. And

moreover, that measure yields $2M$ times $\dfrac{x}{z} + \dfrac{x^3}{3z^3} + \dfrac{x^5}{5z^5}$ &c. From this the following corollary is evident.

Corollary 5

If the sum of two quantities is z and their difference x; and it is assumed that $2M\dfrac{x}{z} = A$, $A\dfrac{xx}{zz} = B$, $B\dfrac{xx}{zz} = C$, $C\dfrac{xx}{zz} = D$ &c, the measure of the ratio which the larger quantity has to the smaller will be $A + \dfrac{B}{3} + \dfrac{C}{5} + \dfrac{D}{7} + $ &c.

Scholium II

By a similar calculation the measure of the ratio between $1 + v$ and 1 will be $M \times \left(v - \dfrac{v^2}{2} + \dfrac{v^3}{3} - \dfrac{v^4}{4} + \dfrac{v^5}{5} - \text{&c:} \right)$. Whence if that measure is called m,

$$\frac{m}{M} = v - \frac{v^2}{2} + \frac{v^3}{3} - \frac{v^4}{4} + \frac{v^5}{5} - \text{&c:}$$

and hence,

$$\frac{mm}{MM} = vv - v^3 + \frac{11}{12}v^4 - \frac{5}{6}v^5, \text{&c;}$$

similarly

$$\frac{m^3}{M^3} = v^3 - \frac{3}{2}v^4 + \frac{7}{4}v^5, \text{&c;}$$

and

$$\frac{m^4}{M^4} = v^4 - 2v^5, \text{&c;}$$

and finally

$$\frac{m^5}{M^5} = v^5, \text{&c.}$$

Thus in turn, from the given measure m, is found the ratio which is measured. By adding like to like we shall discover

$$\frac{m}{M} + \frac{mm}{2MM} = v * - \frac{v^3}{6} + \frac{5}{24}v^4 - \frac{13}{60}v^5, \text{&c;}$$

and again

$$\frac{m}{M}+\frac{mm}{2MM}+\frac{m^3}{6M^3}=v**-\frac{v^4}{24}+\frac{3}{40}v^5,\&c;$$

and again

$$\frac{m}{M}+\frac{mm}{2MM}+\frac{m^3}{6M^3}+\frac{m^4}{24M^4}=v***-\frac{v^5}{120},\&c;$$

and at length

$$\frac{m}{M}+\frac{mm}{2MM}+\frac{m^3}{6M^3}+\frac{m^4}{24M^4}+\frac{m^5}{120M^5}+\&c=v****,\&c.$$

Thus the ratio required, between $1+v$ and 1 is that which $1+\dfrac{m}{M}+$ $\dfrac{mm}{2MM}+\dfrac{m^3}{6M^3}+\dfrac{m^4}{24M^4}+\dfrac{m^5}{120M^5}+\&c$ has to 1. Put $m=M$ or $\dfrac{m}{M}=1$, and thence the ratio modularis will be that which $1+\dfrac{1}{1}+\dfrac{1}{2}+\dfrac{1}{6}+\dfrac{1}{24}+\dfrac{1}{120}$ $\&c$ has to 1. In the same way, if the ratio between 1 and $1-v$ is given, the measure of the ratio will be $M\times\left(v+\dfrac{v^2}{2}+\dfrac{v^3}{3}+\dfrac{v^4}{4}+\dfrac{v^5}{5},\&c.\right)$. And in turn if the measure m of the ratio is given, the ratio will be that which 1 has to $1-\dfrac{m}{M}+\dfrac{mm}{2MM}-\dfrac{m^3}{6M^3}+\dfrac{m^4}{24M^4}-\dfrac{m^5}{120M^5}+\&c$. If m is put equal to M, or $\dfrac{m}{M}=1$, the modular ratio will hence be that which 1 has to $1-\dfrac{1}{1}+\dfrac{1}{2}-\dfrac{1}{6}+\dfrac{1}{24}-\dfrac{1}{120}+\&c$. From hence the following corollary is clear.

Corollary 6
If we define a term R, and assume that $\dfrac{1}{1}R=A$, $\dfrac{1}{2}A=B$, $\dfrac{1}{3}B=C$, $\dfrac{1}{4}C=D$, $\dfrac{1}{5}D=E$, $\&c$ to infinity, and S is taken $=R+A+B+C+D+E+\&c$: the modular ratio will be that which the minor term R (defined) has to the major term S (found). Or, when the term S is defined, if it is assumed that $\dfrac{S}{1}=A$, $\dfrac{A}{2}=B$, $\dfrac{B}{3}=C$, $\dfrac{C}{4}=D$, $\dfrac{D}{5}=E$, $\&c$ to infinity, and R is taken $=S-A+B-C+D-E+\&c$: the modular ratio will be that between the major term S (defined) and the minor term

R (found). Further the same ratio is between 2.718 281 828 459 &c and 1, or between 1 and 0.367 879 441 171 &c.

Scholium III

If it is desired to obtain the minor terms which show the same modular ratio so nearly, that none can be nearer unless they are greater than these, the work will be arranged as follows.

Ratios greater than the true one		Ratios less than the true one	
1	0×2	0	1
2	1	2	0
3	1×2	2	1×1
8	3	6	2
11	4×1	8	3×1
76	28	11	4
87	32×1	19	7×4
106	39	87	32
193	71×6	106	39×1
1264	465	1158	426
1457	536×1	1264	465×1
21768	8008	1457	536
23225	8544×1	2721	1001×8
25946	9545	23225	8544
49171	18089×10	25946	9545×1
&c.	&c.	&c.	&c.

The major term 2.718 28 &c. should be divided by the minor 1, or again the major 1 by the minor 0.367 879 &c. and once more the minor by the number which is left, and this again by the last remainder, and so continue forward; and the quotients 2, 1, 2, 1, 1, 4, 1, 1, 6, 1, 1, 8, 1, 1, 10, 1, 1, 12, 1, 1, 14, 1, 1, 16, 1, 1 &c. will be produced.

Having made these calculations, we must set out two columns of ratios, of which one contains the terms which have a ratio greater than the true one, and the other contains the terms which have a ratio less than the true one; beginning the computation with the ratios 1 to 0 and 0 to 1, which are most remote from the true one, and having started there, continuing to deduce the remaining ratios, which approach ever closer to the true one. Let the terms 1 and 0 be multiplied by the first quotient 2, and let 2 and 0 be written under the terms 0 and 1; and addition will produce the ratio 2+0 to 0+1, or 2 to 1. Let the terms of this [ratio] be multiplied by the second quotient 1, to make 2 and 1 which are added to the terms 1 and 0, and the ratio $2+1$ to $1+0$, or 3 to 1 will be obtained. Let the terms of this (ratio) be multiplied by the third quotient 2 to make 6 and 2 added to the preceding terms 2 and 1; and the ratio 8 to 3 will be obtained. Let the terms of this ratio be multiplied by the fourth quotient 1, to make 8 and 3 which are added to the preceding terms 3 and 1, and the ratio 11 to 4 will be obtained. Let the terms of this ratio be multiplied by the fifth quotient 1, to make 11 and 4 to be added to the preceding 8 and 3, and the ratio 19 to 7 will be obtained. Let the terms of this ratio in the same way be multiplied by the sixth quotient 4, making 76 and 28 to be added to the preceding 11 and 4, to produce the ratio 87 to 32, and so continue forward as far as you wish, transferring alternate factors to alternate columns. When this is done, the ratios greater than the true one will be obtained, 3 to 1, 11 to 4, 87 to 32, 193 to 71, 1457 to 536, 23 225 to 8544, 49 171 to 18 089, &c. The ratios less than the true one will be, 2 to 1, 8 to 3, 19 to 7, 106 to 39, 1264 to 465, 2721 to 1001, 25 946 to 9545, &c., and these are the principal and primary ratios by which the proposed ratio is continually approached.

But if we require the entire series of all the ratios greater than the true one, which are such that no ratio greater than the true one, which comes closer to the true one, can be designated in lesser terms, and if we likewise require the entire series of all the ratios less than the true one, which are such that no ratio less than the true one, which comes closer to the true one, can be designated in lesser terms: then besides those primary ratios which we have just discovered we can obtain other secondary ratios. These occur when the quotient is greater than one. These are found by replacing the multiplication which occurred above by the quotient, by repeated addition of terms, as many times as there are units in the quotient. Thus because the first term was 2, terms 1 and 0 must be added twice to terms 0 and 1, and the sums will give ratios 1 to 1, 2 to 1. These last terms, 2 and 1, because the second quotient was 1, must be added once to terms 1 and 0 and the sums will give the ratio 3 to 1. These terms 3 and 1, because the third quotient was 2, are added twice to terms 2 and 1, and the sums will give the ratios 5 to 2, 8 to 3. These last terms 8 and 3, because the fourth quotient was 1, must be added once to the terms 3 and 1 and the sums will give the ratio 11 to 4. These terms 11 and 4, because the fifth quotient was 1, must be added once to the terms 8 and 3, and the sums will give the ratio 19 to 7. Finally these terms 19 and 7, because the sixth quotient was 4, are added four times to the terms 11 and 4, and the sums will give the ratios 30 to 11, 49 to

18, 68 to 25, 87 to 32. And so one can proceed as far as seems convenient. When this operation is at length completed, the entire series of all the major ratios greater than the true one will be 1 to 0, 3 to 1, 11 to 4, 30 to 11, 49 to 18, 68 to 25, 87 to 32, &c. similarly the entire series of all the ratios less than the true one will be 0 to 1, 1 to 1, 2 to 1, 5 to 2, 8 to 3, 19 to 7 &c.

Ratios greater than the true one		Ratios less than the true one	
1	0×2	0	1
2	1	1	0
3	1×2	1	1
8	3	1	0
11	4×1	2	1×1
19	7	3	1
30	11	5	2
19	7	1	1
49	18	8	3×1
19	7	11	4
68	25	19	7×4
19	7	87	32
87	32×1	106	39×1
&c.	&c.	&c.	&c.

The application of these approximations is widespread, wherefore I have given a somewhat prolix exposition of their invention, by the method which seems to me simplest and easiest. The celebrated men Wallis and Huygens have dealt with the same argument, slightly differently.

PROPOSITION II

To construct the Briggsian canon of logarithms
The logarithms of compound numbers are derived from the logarithms of their prime components, by addition only: the investigation into these can be carried out in a number of ways. I append a single example. By the fifth corollary of the above proposition, writing 1 for M, the logarithms can be found of the ratios between 126 and 125, 225 and 224, 2401 and 2400, 4375 and 4374, which should be called respectively p, q, r and s: and the logarithm of ten will be $239p + 90q - 63r + + 103s$, or 2.302 585 092 994 &c. And so, as the Briggs logarithm of 10 is 1, the result, (by corollary 3, proposition I), would be that the logarithm of 10 which we have just discovered,

2.302 585 092 994 &c., is to its modulus 1, as the Briggs logarithm of 10, 1, is to the Briggs modulus, which will therefore be 0.434 294 481 903 &c. Let this value then be put for M, and $M(202p + 76q - 53r + 87s)$. $M(167p + 63q - 44r + 72s)$, $M(114p + 43q - 30r + 49s)$, will be the Briggs logarithms of the numbers 7, 5, 3. The logarithm of the number 2 is obtained by subtracting the logarithm of the number 5 from the logarithm of the number 10. And thus the Briggs modulus and the logarithms of all the prime numbers less than ten are given. The logarithms of the sequence of prime numbers 11, 13, 17, 19, 23 &c. can thus be computed. First is sought the product of the numbers on either side of the proposed prime, then the prime itself squared, and this always exceeds the product by 1. To the logarithm of the ratio of the square to the product, (using cor. 5 prop. I) should be added the logarithm of the product, (which will always be composed of the given logarithms of primes less than the given prime), and the semi-sum will be the logarithm of the prime, which we are seeking.

Corollary
The modulus of the Briggsian canon is 0.434 294 448 190 3 &c. and the reciprocal of this is 2.302 585 092 994 &c.

   If the logarithm of an intermediate number is required, put a for the intermediate number, and e for the near tabular value, so that a is the larger and e the smaller, their sum z and difference x. Put λ for the logarithm of the ratio of a to e, that is for the excess of the logarithm of a over that of e, and λ will be $2M\dfrac{x}{z}$ approximately. If the number which is congruent to the intermediate logarithm is required, since $\lambda = 2M\dfrac{x}{z} = 2\dfrac{Mx}{2a-x}$ or $2\dfrac{Mx}{2e+x}$; x will be $= \dfrac{\lambda}{M+\frac{1}{2}\lambda}\cdot a$ or $\dfrac{\lambda}{M-\frac{1}{2}\lambda}\cdot e$ approximately.

PROPOSITION III

To show the construction of a canon of logarithms, by any logometric system whatever

Case 1
If the measure of any fixed ratio from the proposed system is given, the measure of any second ratio will be to that given measure of the

fixed ratio as the logarithm of the second ratio is to the logarithm of the same fixed ratio.

Case 2
If a measure of a ratio of the system is not given, then a modulus of the proposed system will have to be found, by the second corollary of the first proposition. And the measure of any (second) ratio will be to the modulus which has been found, as the logarithm of the second ratio to the modulus of the canon.

There are examples of this latter case in the sequel.

PROPOSITION IV

The quadrature of any hyperbolic space by a canon of logarithms
Let *ERSF* be any hyperbola described with centre *A*, asymptotes *ABC*, *AD* [Fig. 2(*a*)]; and let the area *BEFC*, enclosed between the straight lines *BE*, *CF* parallel to the asymptote *AD* be required. Let the parallelogram *ABED* be completed, and the measure of the ratio *AC* to *AB* (or of *BE* to *CF*) be found, (by prop. III). I say the measure will be found equal to the magnitude of the area *BEFC* which we are seeking. Now consider the base of this area as divided into innumerable small particles as small as possible, such as *PQ*, on the condition that wherever the ratio existing between *AQ* and *AP* is given, the lines *PR* and *QS* shall be drawn parallel to the asymptote *AD*. Since also *AQ* is as *AP*, by division *PQ* will be as *AP*, that is, as the reciprocal of *PR*. Hence area *PRSQ* is given, which may accordingly be taken as the measure of the given ratio *AQ* to *AP*. Further, the modulus of this measure will be parallelogram *ABED*, by corollary 2, proposition I. For if the equal parallelogram *APRT* is completed, it will at once be understood that it is to area *PRSQ*, as *AP* is to *PQ*. Therefore, gathering similar calculations of the areas and ratios on each side, the total area *BEFC* will be the measure of the whole ratio *AC* to *AB*, or between *BE* and *CF*, to their modulus *ABED*.

Alternatively. Suppose again a hyperbola (of whatever power) *AP* is described with centre *C* and asymptote *CB*; and the area of any sector *CAP* is required, lying between the semi-diameters *CA*, *CP* and the curve *AP* [Fig. 2(*b*)]. Produce the semi-diameter *CAQ* beyond the vertex *A*. Draw its conjugate *CR*, and let straight lines *PQ* and *PR* be drawn to meet *CQ* and *CR* at right angles from the point *P*, so that they meet the asymptote *CB* at *Z* and *X*. Next draw *AB*, the tangent to the hyperbola at *A*, to cut the asymptote in *B* and the straight line *CP* in *D*: and the triangle *ABC* being the modulus, the area of sector

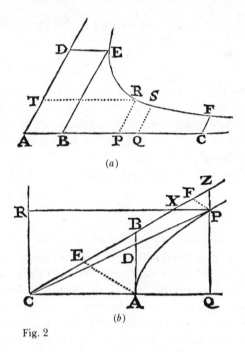

(a)

(b)

Fig. 2

CAP, which we are seeking, will be the measure of the ratio between $QZ + QP$ and AB, or the ratio between AB and $QZ - QP$, or the sub-duplicate ratio between $QZ + QP$ and $QZ - QP$, or the sub-duplicate ratio between $AB + AD$ and $AB - AD$; or it will be the measure of the ratio between $RP + RX$ and CA, or the ratio between CA and $RP - RX$, or the sub-duplicate ratio between $RP + RX$ and $RP - RX$. For if the two straight lines AE, PF which cut the asymptote CB in E and F, and are parallel to the other asymptote, are constructed, all these ratios will be equal to the ratio of AE to PF, or CF to CE; and the sector *CAP* and area *EAPF* will be equal, and similarly triangle *ABC* will be equal to twice the triangle *AEC* or to the parallelogram enclosed by the asymptotes and the hyperbola. So the proposition becomes clear from the above demonstration.

Now area *BEFC* is given by the first method, or area *CAP* by the latter method; similarly any other hyperbolic area will be given, bounded by the arc *EF* or the arc *AP*, since it is always either the sum or the difference of the area just found and some rectilinear area.

Scholium

Hence the solution easily follows of all problems which depend on the quadrature of the hyperbola. A splendid example is afforded by

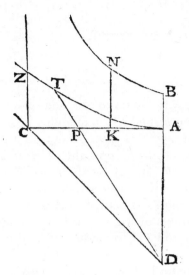

Fig. 3

the descent of heavy bodies in media whose resistance varies as the square of the velocity of a moving body. Let V be the maximum velocity which a body descending infinitely can acquire in this kind of medium, T is the half time for the same body to acquire its velocity in falling through the same medium, under the force of its own relative weight only, apart from the resistance, S the space described in the fall, R the relative weight of the body in the resisting medium; and the space s is required which the descending body traverses in any time t in the resisting medium, also the resistance r which it endures at the end of that time, and the velocity v acquired in its descent.

With centre D, vertex A, let an equilateral hyperbola AT be described [Fig. 3], one of whose asymptotes is DC, and the tangent AC at the vertex is equal to the semi-axis AD. Let the area DAT to half the triangle DAC be taken as t to T, and let DT intersect the tangent AC in P: and v will be to V as AP is to AC. Let AK be a third proportional to AC and AP, and r will be to R as AK is to AC. Let normals CZ, KN, AB be made perpendicular to the tangent AC. With C as centre, and asymptotes CA, CZ, let any hyperbola BN be described: and s will be to S as area $ABNK$ to the rectangle CKN. All these are clear from the eighth and ninth propositions of the second book of Newton's *Philosophia*.

And so accordingly, t is to T as the hyperbolic area DAT is to half the triangle DAC, that is, as half the measure of the ratio between $AC+AP$ and $AC-AP$ to half the modulus of that measure.

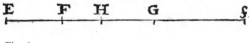

Fig. 4

Therefore, if some straight line *EF* is produced to *f*, and such that *t* is the measure of the ratio between *Ef* and *EF*, to modulus *T*, and *Ff* is bisected in *G*, *GF* will be to *GE* as *AP* is to *AC*, that is as *v* to *V*. Assuming *GE*, *GF*, *GH* in continued proportion, then *GH* and *GE* will be as *AK* to *AC*, that is as *r* to *R* [Fig. 4]. Furthermore, *EG* will be to *EH* as *CA* is to *CK*; and since *s* is to *S* as area *ABNK* to rectangle *CKN*, that is, as the measure of the ratio between *CA* and *CK*, or between *EG* and *EH* to the modulus of the measure, *s* will be the measure of the ratio between *EG* and *EH* to modulus *S*, and hence will be given.

Carrying on from this it is easy to demonstrate, by means of any one hyperbola, an elegant construction, which I have thought fit to add, owing to the importance of the problem. Let a point *F* be taken anywhere between *E* and *G*, and on the other side let *Gf* be taken as equal to *GF*: and let *GE*, *GF*, *GH* be in continued proportion. Then through the points *E*, *F*, *H*, *G*, *f* draw mutually parallel lines *ER*, *FL*, *HM*, *GQ*, *fl*, to be intersected by some hyperbola *LMQl*, which is drawn with centre *E* and asymptotes *ER*, *EG*, and so let the parallelogram *EGQR* be completed. Now if *t* is to *T* as the hyperbolic area *LFfl* is to parallelogram *EQ*, *s* will be to *S* as area *MHGQ* to *EQ*; *v* to *V* as *GF* to *GE*; *r* to *R* as *GH* to *GE*. It is in order to add the other case where the body ascends, lest perhaps that analogy, which ought to hold good for both instances, should seem to fall down to some extent in the construction that is shown. Therefore, with the notation *V*, *R*, *T*, *S*, as before, let *v* and *r* be put for the velocity and resistance under the initial ascent, *s* for the space which the ascending body can describe before the total velocity is lost, *t* for the time of this ascent. *GO* is erected perpendicular to *EG* and equal to *EG* [Fig. 5], and taking points *F* and *f* at the same distances on either side from *G*, let *OF* and *Of* be joined so as to meet in *T* and *t* the circular arc *TGt* described with centre *O*, and let *Gh*, *Gf*, *GE* be in continued proportion, and let *hm* be drawn parallel to *ER*, meeting the hyperbola in *m*. Next, if *t* is the measure of the angle *FOf* to modulus *T*, that is if *t* is to *T* as arc *TGt* to radius *OG*, *s* will be the measure of the ratio between *Eh* and *EG* to modulus *S*, or *s* will be to *S* as the hyperbolic area *mhGQ* to *EQ*, and *v* will be to *V* as *Gf* to *GE*, and *r* to *R* as *Gh* to *GE*.

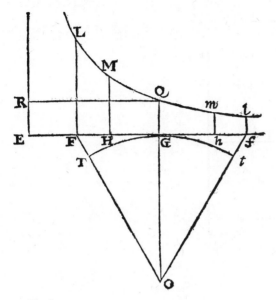

Fig. 5

PROPOSITION V

To describe the logistic curve by a canon of logarithms

If to the asymptote *APCF* of the logistic curve *BQDG*, any two straight lines *AB*, *FG* are drawn at right angles, and include some part *AF* of the asymptote: that part will be the measure of the ratio which the two ordinates have to each other; this is undoubtedly the most noted property of the curve. Therefore a complete and perfect system of logarithms is exhibited by this line: a statement which can also be made of the hyperbola in the previous proposition, and the equiangular spiral in the subsequent; for I omit several other figures which have also long been in use in geometry. Thus, if the position of the asymptote is given, and two points through which the curve must pass, the remaining points will be given by case 1 of proposition III. But if the position of the asymptote is given, and there is also given the modulus of the system and the one point from which to construct the curve, the remaining points will be found by the last case of the same proposition. Indeed it must now be shown how this modulus is to be defined and what importance it has.

Let a straight line *BC* [Fig. 6] be constructed, which touches the curve in *B* and cuts the asymptote in *C*. I say first that the magnitude of the subtangent *AC* is always the same, wherever the point *B* is taken. For the ordinate *PQ* should be understood to be very close to

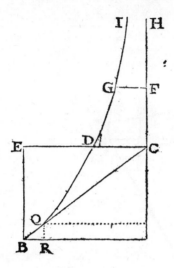

Fig. 6

the ordinate *ARB*, the straight line *QR* parallel to the asymptote *AC*, and the interval of the ordinates, *AP* is given as small as possible. So from the given small line *AP* will be given the ratio of *AB* to *PQ*, and by division, the ratio of *AB* to *RB*, and also (from the properties of the similar triangles *BAC, BRQ*) the ratio of *AC* to *RQ*, or *AP*, and hence the magnitude of *AC* itself.

I say secondly, that this fixed and immutable subtangent *AC* is the modulus, to which all the measures *AF* that are marked off must be determined. That is clear by corollary 2, proposition 1: for as *AB* and *PQ* approach close to equality, *AC* will be to *AP*, which measures the ratio of *AB* to *PQ*, as the term *AB* to the difference of the terms, *BR*. Hence given the subtangent, the description of the curve is easy, and the solution of all problems which depend on it.

If the curve now described is given, the magnitude of the subtangent may be found as follows. Produce some ordinate *CD* to *E*, such that *CE* to *CD* has the ratio of the modulus, defined by corollary 6, proposition 1: and the straight line *EB* drawn from *E* parallel to the asymptote to meet the curve in *B*, will be equal to the subtangent sought.

Corollary 1

The area *ABIH* between the curve *BDI* and the asymptote *ACH*, infinitely extended towards *HI*, and in the other direction bounded by the ordinate *AB*, is equal to the parallelogram *ABEC* formed by the ordinate AB and the subtangent AC. For the area and the parallelogram are composed of elements which are as $AP \times AB$ and $AC \times RB$, which are equated because of the analogy between AP and RB, AC and AB.

Corollary 2
And hence, from the given magnitude of the subtangent, that indefinite area will be as the ordinate which bounds it.

Scholium
The application of this proposition will be made clear by an example. Let it be our intention to discover the density of the atmosphere at any given altitude above the surface of the earth. Let AB be the surface of the earth, and from there the perpendicular AH is produced upwards, and at single points on this line imagine the ordinates FG are drawn, which are as the density of the air at the points F, and all the ends of the ordinates G will be in the logistic line $BDGI$. This is clear by the second corollary of this proposition. For the indefinite area $FGIH$ is as the quantity or weight of the air above F, and that weight is the force which compresses the air in this place, indeed this force, (as a complex experiment shows) is as the density of the compressed air, FG.

Further, if as many altitudes as you please are taken in arithmetical progression, the densities of the air at these altitudes will be in geometrical progression; and the difference of any two altitudes will be the measure of the ratio between the densities of the air at those altitudes.

If the force of gravity is stopped, so that the air is now considered as compressed by some other force, so that it has everywhere the same density as at the surface of the earth; that quantity which was just shown by the area $HABI$ will now be shown by the equal rectangle $ABEC$. The altitude AC of this homogeneous atmosphere is to the height of the mercury in the Torricellian tube, as the weight of mercury is to the weight of air, and hence it is given. Moreover, this given altitude is equal (by corollary 1) to the subtangent of the curve $BDGI$, and also the modulus of the system of all measures AF. Therefore the logarithm of the ratio between the densities of the air at any two altitudes, is to the modulus of the canon, as the difference between the altitudes themselves is to the assumed height AC of the stated homogeneous atmosphere.

The force of gravity at any altitude is thus obtained from the hypothesis. But the *Philosophia* of Newton states that this diminishes

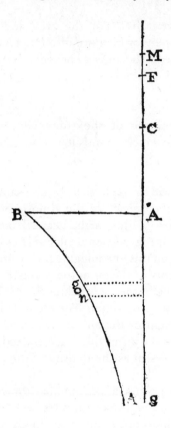

Fig. 7

on receding from the centre of the earth, as the square of the distance. The conclusion therefore will be slightly different. Let *S* [Fig. 7] be the centre of the earth, and *AB* the surface, and let *Sf* be taken as a third proportional to *SF*, *SA*. The ordinate *fg* is erected, which is as the density of the air at *F*, and the curve *Bgn* which always touches the point *g*, will be the same as the former logistica, but in the inverse position. Let the altitude *AF* be increased by the smallest possible particle *FM*, and let *Sm* be taken to *SA* as *SA* is to *SM*. Let the ordinate *mn* be constructed, which is as the density of the air at *M*, and *Sm* will be to *Sf* as *SF* is to *SM*, and by division *fm* to *FM* as *Sf* to *SM*, or as *Sf* to *SF*, that is as SA^2 to SF^2. Whence *fm* is as SF^2 inversely and *FM* directly, that is, as gravitation, and the mass of the air between *F* and *M* conjointly. So *fm* × *fg*, or area *fgmn* is as gravity, quantity and density of the air conjointly, that is as the pressure of that air against the air below, and the sum of all such areas below *fg* is as the sum of all the pressures above F, that is as the density fg of the air at F: and the difference $fgnm$ of the sums as the difference of the densities $fg - mn$. Suppose a small line fm is given, and fg will be as the area $fgnm$, and so as $fg - mn$ and hence (by dividing) as mn. Therefore the given small line fm will be the measure of that given ratio which exists between them, and hence the curve Bgn is clearly the logistic curve. In the same way as with the above described logistic, it easily follows that ordinates very close to the base AB, and disposed at the smallest possible equal intervals, are respectively equal in both curves, and hence the curvature is the same, the inclination of the tangent at point B is the same, and the size of the subtangent is the same.

Therefore, if the distances SF from the centre of the earth are given in musical progression, their reciprocals, namely the distances Sf will be in arithmetical progression, and the densities of the air, fg, will be in geometrical progression.

When investigating therefore the density in some place F, the altitude AF is to be diminished in the ratio of the distance SF to the semi-diameter of the earth SA, and the logarithm of the ratio between the densities of the air at A and F will be to the modulus of the canon, as the diminished altitude Af to the altitude AC of the homogeneous atmosphere.

The demonstrations above hold accurately only if the atmosphere consists entirely of a uniformly elastic air: the ratios which we have mentioned will therefore be a little disturbed if vapours and exhalations are admitted which cause different mixtures of heat and cold to mount to different altitudes.

PROPOSITION VI

To adapt a canon of logarithms to the equiangular spiral
The curved line ADE which is drawn about a pole P, and cuts the radii from the pole, PA, PD, PE &c. at the same given angle [Fig. 8], is called the equiangular spiral. With centre P and any distance PA a circle ABC is described, which meets the radii PA, PD, PE in ABC: I say the arc intercept BC is the measure of the ratio of PD to P and the arc intercept AB is the measure of the ratio of PA to PD. For suppose the arc AB is divided into the smallest possible equal particles, such as QR, and PQ, PR are joined, cutting the spiral in S and T in the given angles PST, PTS: and from the given particle QR the angle QPR will be given, and also the form of the figure SPT, and the ratio of the sides PS, PT. Therefore the given particle QR will be the

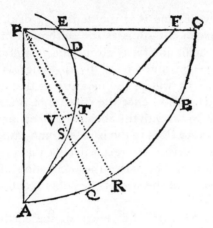

Fig. 8

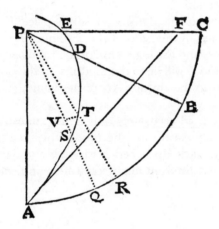

Fig. 9

measure of the ratio of *PS* to *PT*, and the sum of the particles, namely the arc *AB*, will be the measure of a similar sum of ratios, that is of the ratio *PA* to *PD*. And by the same argument, the arc *BC* will be the measure of the ratio of *PD* to *PE*.

Let *AF* [Fig. 9] be drawn touching the spiral at the intersection *A* of the circle and spiral, to be met at *F* by the straight line *PC* which is erected normal to the radius *PA*: and the subtangent *PF* will be the modulus of the measures, by corollary 2, proposition I. For if in the straight line *PS*, *PV* is taken equal to *PT*, and points *VT* are joined, the triangles *PAF*, *VST* will be similar. Whence *PF* is to *VT* as *PA* to *VS*. But *VT* is also to *QR* as *PT* to *PA*: therefore, from mixing the equalities, *PF* is to *QR*, which is the measure of the ratio between *PS* and *PT*, as the side *PT* to the difference of the sides, *VS*.

Scholium

The most illustrious geometer Edmund Halley has happily applied the equiangular spiral to demonstrating the division of the nautical meridian. Let *acp* be the octant of the terrestrial sphere, *p* the pole, *ac* a quadrant of the equator, *ap* a quadrant of the meridian: and the magnitude is sought of the straight line which represents any proposed arc in the planisphere. From the intersection of the equator and the meridian, let a helical line *ade* be understood, which cuts all meridians at a semi right-angle, and is met at *d* by the circle *gd* parallel to the equator, and let the meridian *pdb* be drawn through this point *d*; and the arc length *ab* intercepted on the equator, will be the nautical length of the arc *ag* which we are seeking. Resolve then the arc *ag* into innumerable particles as small as possible such as *gk*, and let the parallel *kmn* be constructed, cutting the meridian *pdb* in *n*, the line *ade* in *m*; and the meridian *pmh* which is drawn will cut off from the equator the particle *bh*, which will be to *mn*, or to *dn* or *gk* which are equal to it (because of the semi right-angle *mdn*) as the periphery of the equator to the periphery of the parallel *kmn*. Therefore the particle *bh* is the nautical magnitude of the particle *gk*, and the sum of all the particles *bh*, namely the length of the arc *ab* is the nautical magnitude of the sum of all the particles *gk*, that is, arc *ag*. Keeping now the equator *abc*, or *ABC* fixed in size, let the surface of the sphere be conceived as depicted stereographically in its (the equator's) plane, and the pole *p* occupying the centre *P* [Fig. 10], the meridians *pga*, *pdb*, *pec* are projected into the same number of straight lines, *PA*, *PDB*, *PEC* radiating from the centre *P*, so that the distance from it of any point *D* or *A* will be the tangent of half the arc *pd* or *pa* which that distance represents. Now the helical line *ade* becomes the equiangular spiral *ADE*, described about the pole *P*, and cutting all its radii at a semi right-angle. The characteristic property of this projection requires that all the angles in the plane and in the spherical surface should remain the same size. Therefore the nautical magnitude *ab* or *AB* of the proposed arc *ag* is to the subtangent *PF*, or to the radius of the sphere *PC* which is now equal to this, as the logarithm of the ratio between *PA* and *PD*, that is between the tangents of the halved arcs *pa* and *pd*, or *pa* and *pg*, to the modulus of the canon.

Hence since the length of the radius is to the length of the arc of one first minute, as 3 437.746 770 784 939 &c. to 1 and the reciprocal

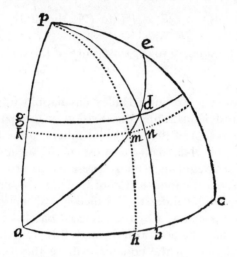

Fig. 10

of the modulus of the canon is 2.302 585 092 994 &c. and these numbers when multiplied make 7 915. 704 467 897 819 &c.: if that nautical length *AB* is to be shown, in first minutes, in the manner required: subtract the artificial tangent of the halved arc *pg* from the artificial tangent of the halved arc *pa*, multiply the result by the number 7 915.704 467 897 819 &c. and the product will give the meridional parts required. A similar conclusion will be reached, whether the point *a* is placed in the equator or in some other place.

SCHOLIUM GENERALE

I have written the preceding pages chiefly for the purpose of showing, using a number of examples, by what a convenient method the use of logarithms can be adapted to geometry, and applied to the solution of difficult problems. At this stage I thought it desirable to add several further constructions, using the ideas so far developed, which came readily to me as I was working on this, so that thus from a wealth of examples, the superiority of the logarithmic method is justified.

In the parabola of Apollonius *AP* [Fig. 11], let *A* be the vertex, *F* the focus and *AQ* the axis, *PQ* the ordinate applied to the axis. Let *AL* be constructed to bisect *PQ* in *L*, and to this produced let *LM* be added, which is the measure of the ratio between *LA + AQ* and *QL*, to modulus *AF*: and the straight line *AM* will be equal to the parabolic arc *AP*.

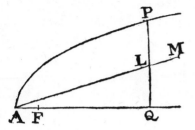

Fig. 11

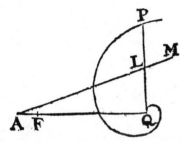

Fig. 12

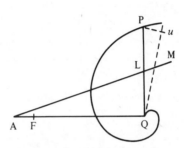

From *Harmonia Mensurarum*
– as amended.

The Archimedean spiral *PQ* [Fig. 12] has a similar extension in a
straight line. Let *Q* be the pole, *QP* the radius constructed from the
pole to cut the curve in some point *P*, and *QA* the normal to this
radius. Construct *LA* parallel to the tangent to the spiral at *P* and
bisecting *QP* in *L*: and putting *AF* to *QL* as *QL* to *QA*, produce *AL*
to *M* so that *LM* is the measure of the ratio between *LA* + *AQ* and
QL, to the modulus *AF*; and the straight line *AM* will be equal to the
spiral arc *PQ*.

Let *A* be the pole of the reciprocal spiral *AeE*, *AB* the infinite prime
radius, *CD* the asymptote parallel to the prime radius and distant *AC*
from it [Fig. 13], and it is proposed to find the length of this curve.
The difference between the common Archimedean spiral, and this,
which I call the reciprocal spiral is that in the former the radii are as
the angles which they make with their prime radius, whereas in the
latter the radii are as the reciprocals of the same angles; in any case
the proportion which the radius *AE* has to the radius *Ae* is the same
as the angle *eAB* has to the angle *EAB*; whence it is easily deduced,
if at points *E* and *e*, tangents *EF* and *eF* are constructed, and *AF*, *Af*
are erected normal to the radii *AE*, *Ae*, these normals will be equal
to each other, and to the asymptotic distance *AC*. Moreover the length

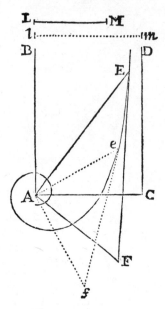

Fig. 13

of any arc *Ee* will be discovered by putting *LM* as the measure of the ratio between *AE* and *EF* – *AF*, to modulus *AF*, and similarly *lm* as the measure of the ratio between *Ae* and *ef* – *Af* to the equal modulus *Af*. For if the difference of the measures *lm* – *LM* is added to the difference of the tangents *EF* – *ef*, the aggregate will be equal to the arc *Ee*.

That logistic line, whose nature we showed to some extent in proposition V, has a not dissimilar determination of its length: which I next show here to the pleasure of those who are delighted by contemplation of this sort. Imagine therefore that the logistic line *EMem* [Fig. 14] is given whose asymptote is *AFaf*: and that the length of any arc *Ee* is required. Drop perpendiculars *ELA*, *ela* to the asymptote and construct the tangents *EF*, *ef*, and let *AL* be taken as equal to the excess by which the tangent *EF* exceeds the subtangent *AF*, and similarly *al* equal to the excess by which the tangent *ef* exceeds the subtangent *af*: and when *LM*, *lm* are drawn parallel to the asymptote, if the difference *lm* – *LM* of the parallels is added to the difference *EF* – *ef* of the tangents, the aggregate will equal the arc *Ee*.

I pass to the cissoid of Diocles [Fig. 15]. Let *A* be its vertex, *AB* the diameter of the generating circle, *BC* the asymptote, *PQ* the perpendicular dropped to the diameter, meeting the cissoid in *P* and the diameter in *Q*. Let *AC* be drawn, which cuts the asymptote in *C* and

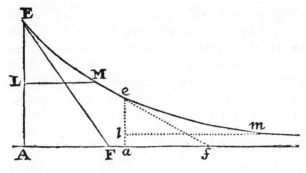

Fig. 14

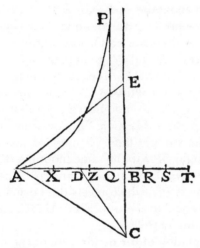

Fig. 15

makes the angle *BAC* which is the third part of a right angle, and
BD is taken as the mean proportional between *BQ* and *BA*, *CD* is
joined. Finally through the mid point of the perpendicular *PQ* construct the straight line *AE*, which meets the asymptote in *E*: and the
arc *AP* of the cissoid will be equal to twice the excess of the straight
line *AE* over the diameter *AB*, together with three times the measure
of the ratio of *BA + AC* to *BD + DC*, to modulus *BC*.

If the cissoidal arc *APQ* is rotated about the axis *AQ*, a solid is
generated, whose dimensions depend upon 'Logometria' and is thus
constructed. Let *AQ*, *AB*, *AR*, *AS*, *AT* be continued proportionals,
whence to modulus *TS*, *QX* is taken as the measure of the ratio
between *AB* and *BQ*, and on the other hand *XZ* is taken equal to *SR*

itself together with one half of *RB* and also one third of *BQ*: and the cissoidal solid having axis *AQ* and semi-diameter of the base *PQ*, will be equal to the cylinder on the same base and whose altitude is *QZ*.

I will add the solid generated from the conchoid of Nichomedes. Let *AE* and *ae* be conjugate curves, drawn with a pole *P*, straight line *CD*, interval *CA* or *Ca*, axis *PaCA* normal to the straight line, and vertices *A* and *a*. From the pole *P* construct any line *PeDE* meeting the straight line in *D*, and the curve in *E* and *e*: and from the nature of the conchoid, the segments *DE*, *De* will be equal to the interval *CA* or *Ca*. With the same interval and centre *P* describe the circular arc *RS* cutting the axis *PC* in *R* and straight line *PD* in *S*: and the semi-sum of the conchoidal solids generated by rotation of the figures *AEDC*, *aeDC* about the axis *AaP* will be to the spherical sector generated by the circular sector *PRS* rotated about the same axis, as $3PC \cdot PD + PR^2$ is to PR^2. Their semi-difference however will be equal to the cylinder whose base is the circle on *Aa* as diameter and whose altitude is twice the measure of the ratio between *PD* and *PC*, to modulus *PC*.

However the area of the total figure *AEea* is equal to the rectangle whose base is *Aa* [Fig. 16], and whose altitude *CM* is the measure of the ratio between *PD* + *DC* and *PC*, to modulus *PC*. So that if it is desired to find the area of the parts *AEDC*, *aeDC*, construct *AF*, *af*, normal to the axis, and in the straight line *CD* must be taken *CN* which is the measure of the angle *CPD* to the same modulus *PC*, and when the straight line *FMf* is drawn through the point *M* to be parallel to the straight line joining points *PN*, which meets the normals in *F* and *f*: the area *AEDC* will be equal to the trapezium *AFMC*, and the area *aeDC* equal to the trapezium *afMC*.

I have given above an explanation of the quadrature of the hyperbola in the way which seemed to me suitable to the proposition. It is agreeable to add another construction here, and at the same time to include the centre of gravity. Let there be shown a part of the interior *ADB* [Fig. 17] included by the curve and some straight line *AB* parallel to the diameter *PQ*. The diameter *CDE* is produced from the figure's centre *C*, which bisects the base in *E*. Hence if in the produced diameter is taken *CR* to *CD*, and *CD* to *CS* as the base *AB* to the diameter *PQ* and to modulus *CS* it makes *CN* the measure of the ratio which *CD* has to *ER*: the triangle *ANE* will be equal to the curvilinear area *ADB*. Moreover, the centre of gravity *Z* of this area will be found by taking *CZ* to *CR* as 2*CR* to 3*EN*.

Suppose [Fig. 18] there is now shown an exterior portion *APQB* enclosed between the opposing curves *AP*, *BQ*, the diameter *PQ*, and any straight line *AB* parallel to the diameter. Let *CD* be the length

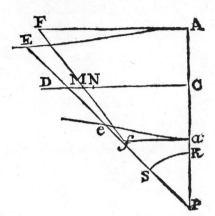

Fig. 16

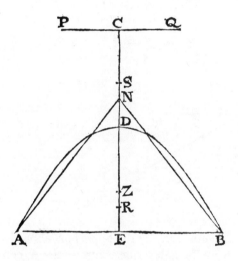

Fig. 17

of the semi-diameter conjugate (to *PQ*) which falls outside the said portion *APQB*, and which produced beyond the centre *C* in the opposite direction bisects the base *AB* in *E*. Hence, if in the diameter produced, *CR* is taken to *CD*, and *CD* to *CS*, and *CS* to *CT*, as the base *AB* to the diameter *PQ*, *CR* and *CT* being on the same side of the centre as the base *AB*: and to the modulus *CS*, on the opposite side of the centre, *CN* is taken as the measure of the ratio which *CD* has to *ER*, the rectilinear triangle *ANB* will be equal to the curvilinear area *APQB*. Moreover, the centre of gravity *Z* of this area will be found by taking *CZ* to *CR* as 2*TR* to 3*EN*.

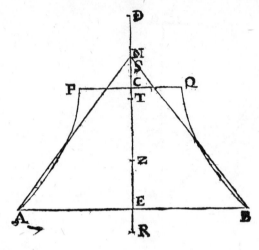

Fig. 18

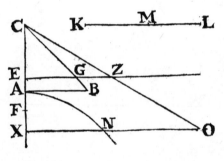

Fig. 19

I proceed to the surface generated by the hyperbola about its axes. Let *AN* [Fig. 19] be the hyperbola described with vertex *A*, centre *C*, asymptote *CB*, focus *F*, principal semi-axis *AC*, conjugate semi-axis *AB* normal to *AC*; at some point *X* in the axis *AC* the ordinate *XN* is drawn, which meets the hyperbola in *N*. In the axis *CA*, *CE* is taken to *CA* as *CA* to *CF*; and having drawn on the same axis the perpendicular *EZ*, which meets the asymptote at *G*, with the angle *CEZ* let a straight line *CZ* be drawn equal to *CX*, which when produced cuts the ordinate *XN* at *O*. Then *KL* is taken equal to the excess of *XO* over *AB*, and *LM* as the measure of the ratio between *CZ+ZE* and *CG+GE* to modulus *CE*: and the surface generated by the rotation of the arc *AN* about the axis *AX* will be to the circle described on *AB* as semi-diameter, as the excess *KM* of *KL* over *LM*, to the semi-diameter *AB*.

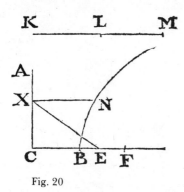

Fig. 20

In the same way let a hyperbola *BN* [Fig. 20] be described with vertex *B*, centre *C*, focus *F*, principal semi-axis *CB*, conjugate semi-axis *CA* normal to *CB*; and at some point *X* in the axis *AC* let the ordinate *XN* be drawn, meeting the hyperbola in *N*. In the axis *CB*, let *CE* be taken to *CA* as *CA* is to *CF*, and let *EX* be joined. Then let *KL* be taken which is to *XC* as *XE* is to *CE*, and *LM*, which is the measure of the ratio between *EX* + *XC* and *CE* to modulus *CE*: and the surface generated by the rotation of the arc *BN* about the axis *CX*, will be to the circle described on *CB* as semi-diameter, as the sum *KM* of the lines *KL* and *LM* to the semi-diameter *CB*.

It will be allowable to add here the surfaces generated by an ellipse. Let *ANB* be the ellipse described with centre *C*, vertices *A* and *B*, focus *F*, principal semi-axis *CB*, conjugate semi-axis *CA*, and at some point *X* in the axis *CA* let the ordinate *XN* be drawn, which meets the ellipse in *N* [Fig. 21]. In the axis *CB* let *CE* be taken to *CA* as *CA* is to *CF*, and let *EX* be joined. Then let *KL* be taken which is to *XC* as *XE* to *CE*, and *LM* which is the measure of ratio between *EX* + *XC* and *CE*, to modulus *CE*: and the surface generated by the arc *BN* rotated about the axis *CX* will be to the circle on *CB* as semi-diameter as the sum *KM* of the lines *KL* and *LM* to the semi-diameter *CB*. In order for this last construction to exist, the semi-axis *CA* about which rotation is made, must be less than the other semi-axis *CB*, for otherwise the quantity of the modulus *CE*, $\dfrac{CA^2}{\sqrt{CB^2 - CA^2}}$ will become impossible, and the logometric construction (which generally arises in this sort of case) becomes trigonometrical, such as that which now follows.

Let *ANB* [Fig. 22] be the ellipse described with centre *C*, vertices *A* and *B*, focus *F*, principal semi-axis *CA*, conjugate semi-axis *CB*; and at some point *X* in the axis *CA* the ordinate *XN* is drawn, which

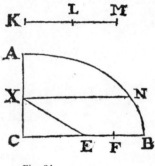

Fig. 21

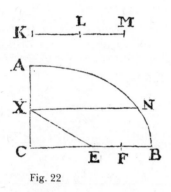

Fig. 22

meets the ellipse in *N*. With angle *CXN* let the straight line *CE* be drawn which is to *CA* as *CA* is to *CF*. Then *KL* is taken which is to *XC* as *XE* to *CE*, and *LM* which is the measure of the angle *XEC* to modulus *CE*, i.e., which is equal to the arc whose sine is *XC* with radius *CE*: and the surface generated by the arc *BN* rotated about the axis *CX*, will be to the circle described on *CB* as semi-diameter, as *KM*, the sum of the lines *KL* and *LM*, to the semi-diameter *CB*. It would have been possible to define this surface by logarithms, but by an impracticable method. For if some arc of a quadrant of a circle described with radius *CE* has sine *CX* [Fig. 23] and sine of the complement of the quadrant *XE*, taking radius *CE* as modulus, the arc will be the measure of the ratio between $EX + XC\sqrt{-1}$ and *CE*, the measure having been multiplied by $\sqrt{-1}$, but I leave this to be examined in more detail by others who will think it worthwhile. Moreover, from the foregoing can be understood the extent of the relationship between the measures of angles and of ratios, further, by simple exchange among themselves, they are easily converted for different cases of the same problem. It has long been observed by

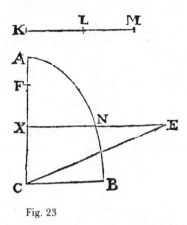

Fig. 23

analysts, that the roots of cubic equations can either be found by Cardan's rules, and so by the discovery of two mean proportionals, or by division of the circular arc into three equal parts if they are hard to explain by the familiar rules [Case 1, Case 2, Case 3]. Descartes observes this, and before Descartes, Francis Vieta observed the same under the section Supplementary Geometry. From hence it is clearly comprehended, to what extent transferring from ratio trisection to angle trisection is the order of nature.

I will be pleased further to declare that marvellous harmony, by an example chosen from the same figure rotated about its own axes. Therefore let $APBQ$ be an ellipse, major axis AB, minor PQ, centre C, focus F. This rotated about either axis generates a solid, whose particles of homogeneous material have a force of attraction in the inverse ratio of the squares of distances; and the power is sought with which that solid attracts some body, placed in that surface, at the extremity of an axis.

The second term being removed, the equation will have three cases. These are resolved by the aid of the triangle ABC [Fig. 24], right angle at A, and in which two sides are always given.

Case 1
If $x^3 + 3a^2x = \pm2a^2b$, AB is put $= a$, $AC = b$; M and N are taken as two mean proportionals between $BC + AC$, and $BC - AC$; $M - N$ will be the unique possible positive root if we have $+2a^2b$ and $N - M$ the unique negative root if we have $-2a^2b$.

Case 2
$x^3 - 3a^2x = \pm2a^2b$, a less than b: put $AB = a$, $BC = b$, and M and N are taken as two mean proportionals between $BC + AC$, and $BC - AC$:

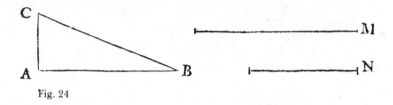

Fig. 24

$M + N$ will be the unique positive root if we have $+2a^2b$, or $-M - N$ the unique possible negative root, if we have $-2a^2b$.

Case 3

If $x^3 - 3a^2x = \pm 2a^2b$, a being greater than b: put $AB = b$, $BC = a$; and M is taken as the sine of one third of the sum of the angles A and B, and N as the sine of one third of the difference $A - B$, the radius being $2BC$: and $-M$, $-N$, $M + N$ will be the three possible roots if we have $+2a^2b$, or M, N, $-M - N$ if we have $-2a^2b$.

And so the solution of all solid problems is obtained, either by a canon of logarithms, or by a trigonometrical canon. Let points P and F [Fig. 25] be joined, and CD be taken as the measure of the ratio between $PF + FC$ and CP, to modulus CA and at the same time CE is taken as the measure of the angle CPF to modulus CP; and let FD be the excess of the measure of CD over CF and FE the excess of CF itself over the measure of CE, and the attraction of the solid generated by rotation about the major axis AB on a body located at A, will be to the attraction on the same body of a homogeneous sphere described about the same axis as $3FD \times CP^2$ to CP^3; again, the attraction of the solid generated by rotation about the minor axis PQ, on a body placed at P will be to the attraction of a homogeneous sphere described about the same axis, on the same body, as $3FE \times CA^2$ to CP^3. Whence the attraction of the first sphere on the body at A is to the attraction of the second sphere on the body at P, as CA to CP: the attraction of the first solid on the body at A to the attraction of the second solid on the body at P will be as $FD \times CP$ to $FE \times CA$.

Hence, because the latter solid is a mean proportional between the first solid and the first sphere, the power of the latter solid on the body at A will be approximately a mean proportional between the power of the first solid and the first sphere, on the same body at A, if the axes of the ellipses are nearly equal. Accordingly, in this case, placing CG as the mean proportional between CF and $3FD$, and taking CH to $3FE$ as CA to CF, the attractions of the latter solid at A and P, or at B and Q will be mutually very closely as CG to CH. This is a useful short cut in enquiring into the shape of the earth, so

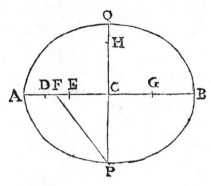

Fig. 25

finely set out by the most celebrated Newton, that most great restorer of health to philosophy.

The consideration of centripetal force presents me with another further example, which yields a sufficiently ample number of different cases. It is proposed to enumerate the kinds of trajectories in which bodies can move, acted on by a centripetal force in the inverse ratio of the cube of the distance, when they are projected from a given place, with a given velocity, following a given direction.

Case 1

Let S be the centre of force [Fig. 26], and let the body leave point P in the direction PQ or QP, with the velocity it would acquire by free fall under the same force towards the centre S, from the point C, and its fall describing the height CP. In the given line QPT, perpendiculars SQ, CT are dropped, and with centre S and distance $\sqrt{SQ^2 + QT^2}$ circle RTA is drawn, meeting line SPC in R, then to modulus $\sqrt{SC^2 - SR^2}$ let arc RA be the measure of the ratio between $SR \pm \sqrt{SR^2 - SP^2}$ and SP but let that arc RA and the point Q lie on different sides of the straight line SR; and the point A will be the highest arch (the apse) of the trajectory. Hence therefore the trajectory is given by taking SM equal to $\sqrt{SC^2 - SR^2}$, then in the straight line SA taking some length SD which is less than SA, and to it erecting a perpendicular DE cutting the circle in E, and joining SE. For if on both sides of the point A, a circular arc AR is taken whose length is the measure of the ratio between $SE + ED$ and SD to modulus SM, and in the semi-diameters SR distances SP are taken equal to SD, the points P will be on the trajectory being described. The time however in which the radius SP drawn from the centre to the moving body sweeps out the area SAP, will be as the straight line DE; for the area swept out

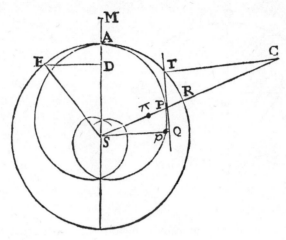

Fig. 26

is equal to *DE* itself × (the halved modulus $\frac{1}{2}SM$). The velocity then of the body in a place such as *P* will be to the velocity with which it could revolve in a circle, radius *SP*, under the same force as $\sqrt{SC^2 - SP^2}$ to *SC*. From its construction it is clear that this first spiral is wound infinitely many times about the centre of force, so as to intersect itself infinitely many times, and all the nodes will be situated on the apsidal line *AS*.

Case 2

Let the point *C* recede to infinity from centre *S*, and let the velocity of the body leaving point *P* along the straight line *PM* or *MP* be that which it would acquire by free fall to the same point *P* from an infinite distance [Fig. 27]. To the straight line *SP* is drawn the normal *SM*, which meets *PM* in *M*; then with centre *S* and radius *SP* let a circle be described, and in its circumference let an arc *PX* be taken whose length is the measure of the ratio between some distance *SD* and the given distance *SP* to modulus *SM*, the arc *PX* and the point *M* lying on different sides of the line *SP*, if *SD* is larger than *SP*, otherwise the same side, and in the radius *SX* let *SZ* be placed equal to *SD*; and the point *Z* will lie on the trajectory being described. The time with which the radius *SZ* drawn from the centre to the moving body sweeps out the area such as *SPZ* will be as $SZ^2 - SP^2$. Now the area swept out is to one half of that difference as half the modulus $(\frac{1}{2}SM)$ is to *SP*. The velocity indeed of the body at point *P* will be the velocity with which it could revolve in a circle with radius *SP*, under the same force. From the construction it is clear that the second spiral is the

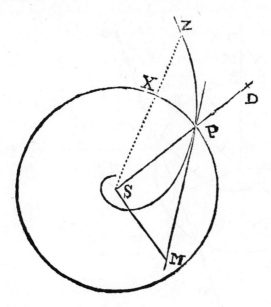

Fig. 27

equiangular spiral of proposition VI, indeed it will become a circle when the angle *SPM* is a right angle.

Case 3

In order that the velocity be greater, point *C* now departs to a distance more than infinite from *S*, or (what is the same thing) approaches to a finite distance on the opposite side of the centre; and the body setting out from point *P* along the line *PQ* or *QP*, its velocity is that which it would acquire in ascending freely from *C* to an infinite distance, and then from an infinite distance on the other side of the centre, descending to *P*, the centripetal force during ascent being equal to the reversed centrifugal force. In the given line *PQT* perpendiculars *SQ*, *CT* are dropped, and *TQ* will be either greater than, equal to, or less than *SQ*. If *TQ* should be greater than *SQ*, with centre *S* and radius $\sqrt{TQ^2 - SQ^2}$ let the circle *RBE* be described meeting the straight line *SP* in *R*, then to modulus $\sqrt{SC^2 - SR^2}$ the arc *RB* is the measure of the ratio between $SR \pm \sqrt{SR^2 + SP^2}$ and *SP*, the arc *RB* and the point *Q* lying on different sides of the straight line *SP*. From this the trajectory is given, taking *SM* equal to $\sqrt{SC^2 - SR^2}$. Taking some length *SD* [Fig. 28] in the straight line *SB*, to it erecting the perpendicular *SE* cutting the circle in *E*, and joining *DE*. Now if in the reverse direction from point *B* the circular arc *BR*

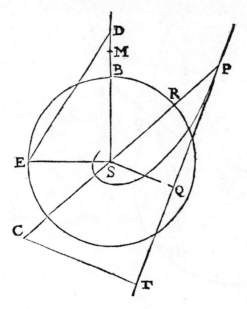

Fig. 28

is taken, whose length is the measure of the ratio between $SE + ED$ and SD to modulus SM, and in the semi-diameter SR a distance SP is taken, equal to SD: the point P will be on the trajectory being described. The time also in which the radius SP drawn from the centre to the moving body sweeps out some area of this trajectory will be as the increment or decrement to the line DE during that time: for the area swept out is equal to this increment or decrement drawn to half the modulus, $\frac{1}{2}SM$. The velocity of the body at a point such as P will be to the velocity with which it could revolve in a circle radius SP under the same force, as $\sqrt{SC^2 + SP^2}$ to SC. From the construction it is clear that this third spiral encircles the centre infinitely many times below the point P, and above that point, either it makes no complete encirclement if the arc RB is less than the whole circumference $RBER$, or it encircles it as many times as the arc exceeds the circumference.

Case 4

Keeping the other factors the same, let TQ and SQ now be equal [Fig. 29]. Let circle PXB be drawn, centre S and radius SP, and let the arc PB be equal to SC, the arc PB and the point Q lying on opposite sides of the straight line SP. From hence the trajectory will be given, taking in the straight line SB some length SD, with centre

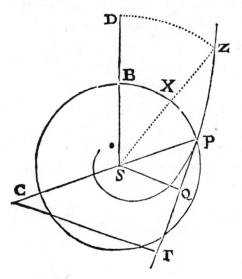

Fig. 29

S and radius SD describing the circular arc DZ equal to SC. For if, curving in opposite directions, arc PB is situated from point P and arc DZ from point D, point Z will be on the trajectory being described. The time in which the radius SZ drawn from the centre to the moving body sweeps out some area SPZ will be as the difference of the radii SZ and SP: for the area swept out will be equal to this constructed difference, by half the distance SC. The velocity of the body at a point such as P, will be to the velocity with which it could revolve in a circle radius SP under the same force, as $\sqrt{SC^2 + SP^2}$ to SC. As is clear from the construction, this fourth spiral is the reciprocal (spiral), the measurement of whose length we have given above.

Case 5

Keeping the other factors the same, let TQ now be less than SQ [Fig. 30]. Let the circle RAE be described, with centre S and radius $\sqrt{SQ^2 - TQ^2}$, meeting the straight line SP in R: then let the arc RA be to the same circular arc whose secant is SP, as $\sqrt{SC^2 + SR^2}$ to SR; then let the arc RA be placed on the same side of the straight line SP as the point Q: and A will be the lowest apse of the trajectory. From hence indeed the trajectory will be given, taking SM equal to $\sqrt{SC^2 + SR^2}$, in the straight line SA taking some length SD which is greater than SA, drawing DE which touches the circle in E, and joining SE. For if on each side of the point A is placed a circular arc AR, whose length is the measure of the angle DSE to modulus SM, and

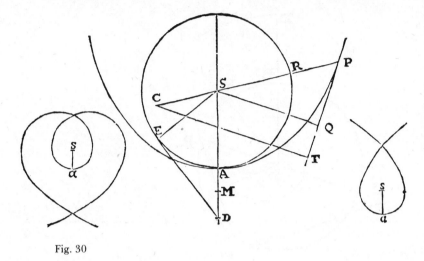

Fig. 30

in the semi-diameters *SR* are taken distances *SP* equal to *SD*, the points *P* will be on the trajectory being described. The time however with which the radius *SP*, drawn from the centre to the moving body, sweeps out some area *SAP*, will be as the straight line *DE*, for the area swept out will be equal to *DE* times half the modulus ($DE \times \frac{1}{2}SM$). The velocity of the body in such a position *P*, will be to the velocity with which it could revolve in a circle of the same radius *SP*, under the same force, as $\sqrt{SC^2 + SP^2}$ to *SC*. From the construction it is clear that this fifth spiral either has no nodes, or one, or several, according to the different proportions of the straight line *SM* to the diameter of the circle *EAR*: the trajectory will intersect itself as often as the straight line exceeds the diameter, and all the nodes will be situated on the apsidal line *AS*.

These therefore are the five species of trajectory. The first and last of these were formerly described by Newton, by the quadrature of the hyperbola and the ellipse.

Geometers will be at liberty, from the examples so far adduced, to judge our method, which if good, they will continue to refine, and in refining, will advance further. There appears indeed to be a most wide field in which they will be able to discover their powers, especially if they might join trigonometry more closely to logometry, which acting together, have a really remarkable affinity, as I have noted before. I would not find it easy to believe that more general principles than these could be given; as the whole of our knowledge contains within it scarcely anything beyond the theory of angles and ratios. Nor will he hope for anything more convenient, who observes the ease of execution as a result of those full and complete numerical tables, not only of logarithms, but also of sines and tangents which, having been received, we owe to the most praised skill of our predecessors. Indeed, so that greater profit may ensue for us, from such a benefaction, it now remains necessary to show by what means the conclusions of the method might be arrived at most advantageously. For this purpose I would have added some theorems, not only of logometry but also of trigonometry, which I keep ready for use; except that it seemed more prudent, since they could not be taught without much obscurity, to pass them by untouched, and leave the investigating yet again to others. But there is not always need for this apparatus for, in the method of fluxions, it often happens that the fluents can be satisfactorily obtained by means of logometry, without the aid of this method, which I will show by one or two examples.

We required from the foregoing, the descent in a straight line of heavy bodies, retarded by the continuous resistance of the medium, from the hypothesis that the resistance was proportional to the square of the velocity. Let it now be proposed, from the same hypothesis, to find the resistance of a suspended body oscillating in a cycloid. The cycloid having accordingly been laid out in a straight line [Fig. 31], let AC be the half, C the lowest point, B the point from which the body begins to fall, BC, CD the arcs the body describes in descent and subsequently in ascent. When these points are established, we must begin by asking what ratio the resistance of the body in a place such as E has to its relative weight in the resisting medium. The weight is shown by AC, and its force derived from the same, by which the pendulum is accelerated at E, will be shown by CE: if this is called x, and its momentum $+\dot{x}$, the momentum for the arc BE described will be $-\dot{x}$. The force of the resistance is denoted by z; and the true force by which the pendulum is accelerated, will be as the excess of the first force over the resistance, that is as $x - z$. Accordingly since the resistance is as the square of the velocity, the momentum $\dot{z}$ of the resistance will be as the velocity and the momentum of the velocity, that is, as $-x$ and $x - z$, or as $z\dot{x} - x\dot{x}$. Now if the time taken is divided into equal particles, the velocity will be as the momentum $-\dot{x}$, of the arc just described, and the momentum of the velocity as the accelerating force $x - z$, which momentum it generates. Since therefore $\dot{z}$ is as $z\dot{x} - x\dot{x}$, if an invariable quantity a is taken, which is of a suitable magnitude, $a\dot{z}$ will equal $z\dot{x} - x\dot{x}$.

Towards the solving of this equation, let a variable quantity x be assumed, and let the equation $z = p + qx + rv$ be formed, in which p, q, r designate other new constant quantities: and $\dot{z}$ will be $= q\dot{x} + r\dot{v}$.

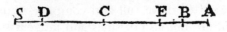

Fig. 31

These same values of z and $\dot{z}$ having been substituted in turn in the equation $a\dot{z} = z\dot{x} - x\dot{x}$, $\overline{aq - p}$, $\dot{x} + ar\dot{v} = \overline{q - 1}$, $x\dot{x} + rv\dot{x}$ will be obtained. To simplify this equation, let $q - 1$ be put equal to 0, $aq - p = 0$; or $q = 1$ and $p = a$: and this will make $\dfrac{a\dot{v}}{x} = \dot{x}$: and also $z = a + x + rv$. With two points D and S taken on the same side of C, let it be understood that CS equals a itself: and z will be $= SE + rv$, and $CS\dot{v}/v = \dot{x}$. Let c be the value of v when the point E is at C: and the quantity x, or CE will be equal to the measure of the ratio which v has to c, to modulus CS, by the first proposition: which equality it is my custom to designate $CE = CS\Big|_{c}^{v}$. All the difficulty of the problem now depends on these two equations, $CE = CS\dfrac{v}{c}$ and $x = SE + rv$. It will not indeed be possible to put them to use, before the quantities r and CS have been determined. For effecting this, two conditions remain, not yet fulfilled. It is necessary for the resistance to be nil, and so the quantity z or $SE + rv$ to vanish, when the point E coincides with points B and D.

Let therefore b and d be the values of v when point E coincides with B and D respectively: and in these cases will be obtained $SB + rb = 0$, $SD + rd = 0$. Whence $r = -\dfrac{SB}{b}$, and $r = -\dfrac{SD}{d}$, and $z = SE + rv = SE - \dfrac{v}{b}SB = SE - \dfrac{v}{d}SD$. Again, $\dfrac{SB}{SD}$ will be $= \dfrac{b}{d}$ and so $CS\dfrac{SB}{SD} = \Big(CS\Big|_{d}^{b} = CS\Big|_{c}^{b} - CS\Big|_{c}^{d} = CB + CD = \Big)BD$: whence point S will be given.

The problem therefore will be completed in this manner. Let BD be produced from D to S, until at length BD be the measure of the ratio between SB and SD to modulus CS. Thence, a free choice of the quantity c having been made, quantities b and v should be taken in such a way that, with the same modulus CS, CB will be the measure of the ratio of b to c, and CE will also be the measure of v to c, and the force of the resistance in the place E, to the relative weight of the suspended body, will be as $SE = \dfrac{v}{b}SB$, to CA.

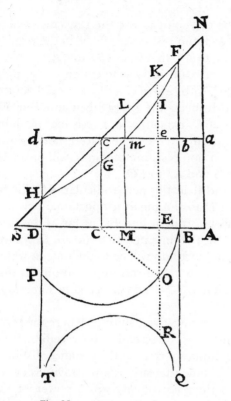

Fig. 32

The solution of this problem has a usefulness in physics, not to be despised. Wherefore it has seemed advisable to add the linear construction of the same, deduced from the same analysis. Point *S* having been found as above, let *DH, Cc, EK, BF, AN* be erected [Fig. 32] perpendicular to the straight line *SA*, to meet the straight line *SN* drawn through *S* in *H* [*a* misprinted for *c*], *K, F, N*. From point *c* the straight line *da* is drawn parallel to the straight line *DA*, which meets these same perpendiculars in *d, c, e, b, a*; and with asymptote *SA* the logistic (curve) *HGIF* is drawn, which passes through points *H* and *F*, and cuts the perpendiculars *Cc, EK* in *G* and *I*, and the parallel *da* in *m*: indeed with these having been placed, the relative weight of the suspended body will be to the force with which that body is accelerated towards the point *E* in a non-resisting medium, as *aN* to *eK*; it will also be to the force of the resistance in the place *E*, as *aN* to *KI*; and also to the force with which the pendulum is accelerated to point *E* in a resisting medium, as *aN* to *eI*. Again, if from point *m* a perpendicular *LmM* is drawn to the straight line *SMA*, which cuts SN in $L:M$ will be the point where the resistance would be maximum, and also that maximum resistance will be to the relative weight of the pendulum, as Lm to Na, that is, as CM to CA.

But if the straight line SN were drawn, so as to cut off the straight line DH which will be twice SD, with centre C and radius CB, let the circle BOP be described, which meets the perpendicular KE in O: the velocity of the pendulum at E oscillating in a resisting medium, to the velocity of the same pendulum in the same place E borne down by the same relative weight in a non resisting medium, will be as the mean proportional between CS and KI to EO.

Further, if CO is joined, and if in the perpendicular KE, ER be assumed, which is to CB as CD to the mean proportional between $4CS$ and KI, and in drawing the straight line ER as far as the base BE the area $BQRE$ is generated: the time in which the arc BE of the cycloid is described in a resisting medium will be to the time in which the same arc would be described in a non-resisting medium, as that area $BQRE$ to the circular sector BOC. I now proceed to something different.

We find the density of the air at any given altitude, where the force of gravity was either uniform, or was decreasing on receding from the centre of the earth in the duplicate ratio of the distance. I should like to discover in addition the same (density) where gravity is either augmented or diminished in the ratio of any given power of the distance. Let S be the centre of the earth [Fig. 33], A a point in its surface or any other place, SAF_2 a straight line produced from the centre of the earth to the top of the atmosphere: and let the ratio be enquired of the density in place A, to the density in some place F, from the hypothesis that the force of gravity at place F is as SP^n, the value of the distance SF wherever it occurs, whose index is n. Let x be written for SF, and let d and v designate the values of the density of the air at A and F; and since the density is everywhere as the total pressure of the incumbent air, the momentum of the density will be as the momentum of the pressure, that is $\dot{v}$ as $v\dot{x}x^n$, and also $\dfrac{\dot{v}}{v}$ as $\dot{x}x^n$. Let AC be the altitude of the atmosphere, if its uniform density were the same as at A; or let AC be to the height of the mercury of the barometer at A, as the density of mercury to the density of the air in the same place A, the height of the mercury in the barometer at A will be to the height of the mercury in the barometer at F, as AC to FC. So the density d of the air at A is to the density v of the air at F as AC to FC: hence it follows that $d-v$ or $\dot{v}$ is to d or v as

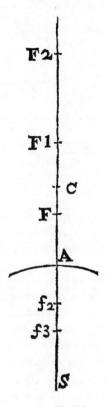

Fig. 33

AF or $\dot{x}$ to AC. Therefore in this case $AC\dfrac{\dot{v}}{v} = \dot{x} = \dfrac{\dot{x}x^n}{SA^n}$. Since there-

fore wherever point F was taken, $\dfrac{\dot{v}}{v}$ was as $\dot{x}x^n$: it will also be true

that $AC\dfrac{\dot{v}}{v} = \dfrac{\dot{x}x^n}{SA^n}$, wherever point F is taken.

Now if gravity is as the reciprocal distance from the centre, or as $\dfrac{1}{x}$ or x^{-1}, n will be = to -1, and thence $AC\dfrac{\dot{v}}{v} = SA\dfrac{\dot{x}}{x}$; hence if the fluents are put equal, the measure of the ratio between the densities d and v to modulus AC, will be equalled by the measure of the ratio between the distances SF and SA to modulus SA.

If the law of gravity should be any other: since $AC\dfrac{\dot{v}}{v}$ is $\dfrac{\dot{x}x^n}{SA^n}$, if the fluents are put equal, $\dfrac{1}{n+1} \times \dfrac{SF^{n+1}}{SA^n} - SA$ will be the measure of the

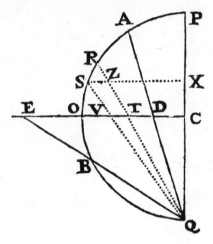

Fig. 34

ratio between the densities d and v, to modulus AC. Therefore, if the increasing terms SA, SF, SF_1, SF_2, etc. are assumed to be in geometric progression; and the decreasing terms SF, SA, Sf_2, Sf_3, etc. the measure of the ratio between the densities of the air at A and F, to modulus AC, will be $\frac{1}{2}Af_3$, if gravity is reciprocally as the cube of the distance; it will be Af_2, if gravity is reciprocally as the square of the distance; it will be AF, if gravity is put uniform; it will be $\frac{1}{2}Af_1$, if gravity is as the distance; it will be $\frac{1}{3}AF_2$, if gravity is as the square of the distance. And so it continues to infinity.

Finally, so that it may be more fully established that synthetic demonstrations may be devised with little trouble from the above principles, it will suffice to add just one more example, since it would be tedious in any case to offer more now. Let us therefore revert to that nautical division of the meridian which we touched on above, and let us see whether, even without the help of any logometric curve, a somewhat similar demonstration can be found in the following manner. Let $PXCQ$ be the axis of the earth, CO the semi-diameter of the equator, $PAOBQ$ the meridian, and the magnitude in the nautical planisphere of any arc AB is to be found. Let straight lines QA, QB be constructed from one of the poles Q or P to the bounds of the arc A and B, meeting the semi-diameter CO in D and E. I say that the nautical magnitude of the arc AB is equal to the measure of the ratio between EC and DC, to modulus OC. For consider the arc AB [Fig. 34] as divided into the smallest possible particles such as RS, and let QR, QS be joined, which cut CO in T and V, and when a perpendicular, SX, to the axis is drawn which meets the straight line

QR in *Z*, the little line *SZ* is equal to the particle *RS*. Therefore the nautical magnitude of the arc *RS* that is produced will be to *OC* the semi-diameter of the sphere, as that arc *RS* or the little line *SZ* to *SX*, that is as *VT* to *VC*. Hence (by corollary 2, proposition 1) the magnitude of that nautical will be equalled by the measure of the ratio between *VC* and *TC*, to modulus *OC*: and similarly, collecting the totals on both sides, the magnitude of the nautical of the total arc *AB* will be equalled by the measure of the total ratio between *EC* and *DC* to the same modulus *OC*.

APPENDIX 2

Commentary on Proposition IV of Logometria

First of all, without reference to the form of the curve, by Proposition I, PQ/AP is the measure of the ratio AQ/AP (Appendix 1, Fig. 2(a)). From the form of the curve, we now have that PR is reciprocally as AP, so that PQ/AP is as $PQ \times PR$. Hence the small element of area $PRSQ$ is a measure of the ratio of AQ to AP. But PQ/AP is also = area $PRSQ$/area $APRT$, and area $APRT$ is constant, being equal to the power of the hyperbola, i.e., c^2 if the equation is taken in the form $xy = c^2$.

Hence, by Proposition I, Corollary 2, area $ABED$ will serve as a modulus for the system of measures of the ratio $AQ/AP = PR/QS$ (from the properties of the curve). This is therefore a logometric system in Cotes' sense: nothing essentially new. It could, however, be pointed out that the curve $xy = 1$ gives the system of logarithms modulus 1, i.e., logarithms base e. If the axes are inclined at $\sin^{-1} 0.434\,294 \ldots$, the Briggs logarithms result.

In the alternative presentation, Cotes makes the system yield substantially more by using the equality of the hyperbolic space $EAPF$ and the sectorial area CAP (Appendix 1, Fig. 2(b))

Note that triangle $OBC = \frac{1}{2}b(a^2/b) = a^2/2$ (Fig. 1). Triangle $OAD = a^2/2$. Therefore area $ABCD$ = area OCD. With the notation of Fig. 2(b) (Appendix 1), where ZP is assumed to meet the hyperbola again in P', we may briefly explain Cotes' string of ratios thus: Area CAP is a measure of $AE:FP = CF:CE = AB:PZ = AB:(QZ - QP)$ and (from the well-known property of the hyperbola $ZP \cdot ZP' = $ constant $= AB^2$) it follows that

$$(QZ - QP) \cdot (QZ + QP) = AB^2,$$

$$\frac{AB}{QZ - QP} = \frac{QZ + QP}{AB} = (\text{say } r),$$

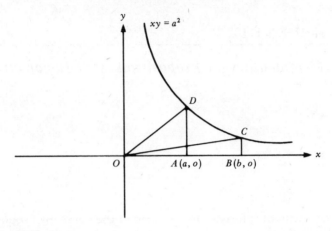

Fig. 1

from which

$$\sqrt{\frac{QZ+QP}{QZ-QP}} = r,$$

and from the similar triangles CAB, CQZ this is equal to

$$\sqrt{\frac{AB+AD}{AB-AD}}.$$

Thus, the area of the hyperbolic sector is the measure of

$$\frac{AE}{FP} = \frac{CF}{CE} = \frac{AB}{QZ-QP} = \frac{QZ+QP}{AB} = \sqrt{\frac{QZ+QP}{QZ-QP}} = \sqrt{\frac{AB+AD}{AB-AD}},$$

the modulus of all these equal measures being triangle CAB. This geometrical presentation can be related to Cotes' tables of integrals which form Part II of Logometria as it was published in *Harmonia Mensurarum*, ed. R. Smith (Cambridge, 1722), but which did not appear in the original publication in the *Philosophical Transactions of the Royal Society*, vol. 29, no. 338 (1714).

From the geometry of the hyperbola and from his Proposition I, Cotes has arrived at logarithmic expressions for the area of a hyperbolic space such as *BEFC* (Appendix 1, Fig. 2(*a*)), or a hyperbolic sector such as *CAP* (Appendix 1, Fig. 2(*b*)). Also, triangle *CQP* − sector *CAP* = hyperbolic space *AQP*. He is now ready to solve any problem of which the hyperbola is a model. Before looking at the illustrative examples it will be instructive to consider these results a little more fully.

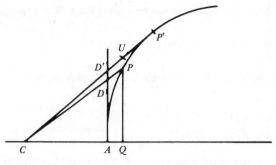

Fig. 2

The result 'sectorial area = measure of the ratio $QZ + QP$ to AB, to modulus $\triangle ABC$', when written in the notation introduced by Cotes (*Harmonia Mensurarum*, p. 37) appears as

$$\text{sectorial area} = \triangle ABC \left| \frac{QZ + QP}{AB} \right. .$$

Using conventional modern notation, and taking the equation of the hyperbola as $(x^2/a^2) - (y^2/a^2) = 1$, we have

$$\text{sectorial area} = \tfrac{1}{2}a^2 \ln \left(\frac{x + \sqrt{x^2 - a^2}}{a} \right)$$
$$= \frac{a^2}{2} \cosh^{-1} \frac{x}{a}$$

(clear recognition of hyperbolic functions was to wait until later in the eighteenth century). Cotes' final expression for the sectorial area is

$$\frac{a^2}{2} \left| \sqrt{\frac{AB + AD}{AB - AD}} \right. .$$

Consider two neighbouring points P, P' on the hyperbola $x^2 - y^2 = a^2$. Join CP, CP', meeting the tangent at A in D, D', and the ordinate at Q in P, U. CPU can be regarded as the element of sectorial area, and expressed in terms of the variable $AD = \text{T}$ (say): (Fig. 2).

$$CPU = \frac{a^3}{2} \frac{\dot{\text{T}}}{a^2 - \text{T}^2}$$

is quickly obtained. Integration of this result gives:

$$\text{sectorial area} = \frac{a^2}{2} \ln \sqrt{\frac{a + \text{T}}{a - \text{T}}},$$

agreeing with Cotes' result.

It is of more interest to look at Cotes' own tables of integrals:

$$\frac{a^3}{2}\frac{\dot{T}}{a^2-T^2} \text{ is Cotes' Form II, } \frac{d\dot{z}\, z^{\theta n+\frac{1}{2}n-1}}{e+fz^n},$$

with $a^3/2$, a^2, -1, 0, 2, T substituted for d, e, f, θ, η, z. If in the expression

$$\frac{a^2}{2}\left|\sqrt{\frac{AB+AD}{AB-AD}}\right.$$

we put $AB = a = R$ the semi-axis, $AD = T$ the tangent at the vertex,

$$\frac{a^2}{2}\left|\sqrt{\frac{AB+AD}{AB-AD}}=\frac{R^2}{2}\right|\sqrt{\frac{R+T}{R-T}}=\frac{R^2}{2}\left|\frac{R+T}{\sqrt{R^2-T^2}}=\frac{R^2}{2}\right|\frac{R+T}{S},$$

where $R^2-T^2=S^2$, and this agrees with the result given by Cotes (*Harmonia Mensurarum*, p. 47) for Form II, $\theta = 0$, i.e.,

$$\frac{2}{\eta e}\,d R\left|\frac{R+T}{S}=\frac{R^2}{2}\right|\frac{R+T}{S}.$$

Cotes' eighteen forms of fluents

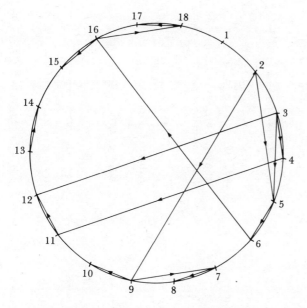

Fig. 1. Diagram illustrating the order in which Saunderson derived the integrals of Cotes' eighteen forms, using substitution, 'Saunderson's rule', and reduction formulae.

No.	Form	Values of θ tabulated in *Harmonia Mensurarum*	Reduction formulae and notes
I	$\dfrac{dz^{\theta n-1}z}{e+fz^n}$	-4 to $+5$	$I_{\theta+1} + \dfrac{e}{f}I_\theta = \dfrac{dz^{\theta n}}{\theta nf}$ for $\theta \geq 1$. The transformation $e, f, \eta \to f, e, -\eta$ transforms I_θ to $I_{1-\theta}$, and can be used for $\theta < 1$.
II	$\dfrac{dz^{\theta n+\frac{1}{2}n-1}z}{e+fz^n}$	-3 to $+3$	$(\theta+\tfrac{1}{2})eI_\theta + (\theta+\tfrac{1}{2})fI_{\theta+1} = \dfrac{d}{\eta}z^{\theta n+\frac{1}{2}n}$.
III	$dz^{\theta n-1}\sqrt{e+fz^n}\,z$	-3 to $+4$	$\theta eI_\theta + (\theta+\tfrac{3}{2})fI_{\theta+1} = \dfrac{d}{\eta}z^{\theta n}(e+fz^n)^{\frac{3}{2}}$. Also $eI_\theta(\mathrm{v}) + fI_{\theta+1}(\mathrm{v}) = I_\theta(\mathrm{III})$, where here, and in the sequel, Roman numerals refer to the relevant fluxional forms.
IV	$dz^{\theta n+\frac{1}{2}n-1}\sqrt{e+fz^n}\,z$	-3 to $+4$	$(\theta+\tfrac{1}{2})eI_\theta + (\theta+\tfrac{5}{2})fI_{\theta+1} = \dfrac{d}{\eta}z^{\theta n+\frac{1}{2}n}(e+fz^n)^{\frac{3}{2}}$. Also $e, f, \eta \to f, e, -\eta$ transforms $I_\theta(\mathrm{III})$ to $I_{-(\theta+1)}$.
V	$\dfrac{dz^{\theta n-1}z}{\sqrt{e+fz^n}}$	-3 to $+4$	$\theta eI_\theta + (\theta+\tfrac{1}{2})fI_{\theta+1} = \dfrac{dz^{\theta n}}{\eta}(e+fz^n)$.
VI	$\dfrac{dz^{\theta n+\frac{1}{2}n-1}z}{\sqrt{e+fz^n}}$	-4 to $+3$	$(\theta+\tfrac{1}{2})eI_\theta + (\theta+1)fI_{\theta+1} = \dfrac{d}{\eta}z^{\theta n+\frac{1}{2}n}(e+fz^n)^{\frac{1}{2}}$. Also $e, f, \eta \to f, e, -\eta$ transforms $I_{-\theta}(\mathrm{VI})$ to $I_\theta(\mathrm{III})$.
VII	$\dfrac{dz^{\theta n-1}\sqrt{e+fz^n}\,z}{g+hz^n}$	-2 to $+3$	$\theta egI_\theta + [(\theta+\tfrac{3}{2})fg + (\theta+\tfrac{1}{2})eh]I_{\theta+1} + (\theta+\tfrac{3}{2}+\tfrac{1}{2})fhI_{\theta+2} = \dfrac{d}{\eta}z^{\theta n}(e+fz^n)^{\frac{3}{2}}(g+hz^n)^{\frac{1}{2}}$. Also $eI_\theta(\mathrm{IX}) + fI_{\theta+1}(\mathrm{IX}) = I_\theta(\mathrm{VII})$.

VIII

$$\frac{dz^{\theta\eta+\frac{1}{2}\eta-1}\sqrt{e+fz^\eta}\,z}{g+hz^\eta}$$

-3 to $+2$

$(\theta+\frac{1}{2})egI_\theta+[(\theta+\frac{5}{2})fg+(\theta+1)eh]I_{\theta+1}+(\theta+\frac{7}{2})fhI_{\theta+2}$
$$=\frac{d}{\eta}z^{\theta\eta+\frac{1}{2}\eta}(e+fz^\eta)^{\frac{3}{2}}(g+hz^\eta)^{\frac{1}{2}}.$$
Also $e,f,g,h,\eta\to f,e,h,g,-\eta$ transforms $I_\theta(\text{VII})$ to $I_{-\theta}(\text{VIII})$.

IX

$$\frac{dz^{\theta\eta+\frac{1}{2}\eta-1}z}{(g+hz^\eta)\sqrt{e+fz^\eta}}$$

-1 to 4

$\theta egI_\theta+[(\theta+\frac{1}{2})fg+\theta eh]I_{\theta+1}+(\theta+\frac{1}{2})fhI_{\theta+2}=\frac{d}{\eta}z^{\theta\eta+\frac{1}{2}\eta}(e+fz^\eta)^{\frac{1}{2}}$. Also
$gI_\theta(\text{IX})+hI_{\theta+1}(\text{IX})=I_\theta(\text{V})$.

X

$$\frac{dz^{\theta\eta+\frac{1}{2}\eta-1}z}{(g+hz^\eta)\sqrt{e+fz^\eta}}$$

-3 to $+2$

$(\theta+\frac{1}{2})\theta gI_\theta+[(\theta+\frac{3}{2})fg+(\theta+\frac{1}{2})eh]I_{\theta+1}+(\theta+1)fhI_{\theta+2}=\frac{d}{\eta}z^{\theta\eta+\frac{1}{2}\eta}(e+fz^\eta)^{\frac{1}{2}}$.
Also $e,f,g,h,\eta\to f,e,h,g,-\eta$ transforms $I_\theta(\text{IX})$ to $I_{1-\theta}(\text{X})$.

XI

$$dz^{\theta\eta-1}\sqrt{\frac{e+fz^\eta}{g+hz^\eta}}\,z$$

-3 to 3

$\theta egI_\theta+[(\theta+\frac{3}{2})fg+(\theta+\frac{1}{2})eh]I_{\theta+1}+(\theta+2)fhI_{\theta+2}=\frac{d}{\eta}z^{\theta\eta}(e+fz^\eta)^{\frac{3}{2}}(g+hz^\eta)^{\frac{1}{2}}$.
Also $e,f,g,h,\eta\to f,e,h,g,-\eta$ transforms $I_\theta(\text{XI})$ to $I_{-\theta}(\text{XI})$.

XII

$$\frac{dz^{\theta\eta-1}\sqrt{\dfrac{e+fz^\eta}{g+hz^\eta}}\,z}{(k+lz^\eta)}$$

-2 to $+3$

$\theta ghI_\theta+[(\theta+\frac{3}{2})fgk+(\theta+\frac{1}{2})ehk+\theta egl]I_{\theta+1}+[(\theta+2)fhk+(\theta+\frac{3}{2})fgl$
$+(\theta+\frac{1}{2})ehl]I_{\theta+2}+(\theta+2)fhlI_{\theta+3}=\frac{d}{\eta}z^{\theta\eta}(e+fz^\eta)^{\frac{3}{2}}(g+hz^\eta)^{\frac{1}{2}}$.
Also $kI_\theta(\text{XII})+lI_{\theta+1}(\text{XII})=I_\theta(\text{XI})$.

XIII

$$\frac{dz^{\theta\eta-1}z}{e+fz^\eta+gz^{2\eta}}$$

-2 to $+4$

$$eI_\theta+fI_{\theta+1}+gI_{\theta+2}=\frac{dz^{\theta\eta}}{\theta\eta}.$$

XIV

$$\frac{dz^{\theta\eta-1}z}{(k+lz^\eta)(e+fz^\eta+gz^{2\eta})}$$

-1 to $+4$

$$ekI_\theta+(fk+el)I_{\theta+1}+(gk+fl)I_{\theta+2}+glI_{\theta+3}=\frac{dz^{\theta\eta}}{\theta\eta}.$$
Also $kI_\theta(\text{XIV})+lI_{\theta+1}(\text{XIV})=I_\theta(\text{XIII})$.

No.	Form	Values of θ tabulated in *Harmonia Mensurarum*	Reduction formulae and notes
XV	$dz^{\theta\eta-1}\sqrt{e+fz+gz^2}\,i$	−4 to +3	$\theta eI_\theta+(\theta+\frac{3}{2})fI_{\theta+1}+(\theta+3)gI_{\theta+2}=\dfrac{dz^{\theta\eta}}{\eta}(e+fz^\eta+gz^{2\eta})^{\frac{3}{2}}$. Also $eI_\theta(\text{XVI})+fI_{\theta+1}(\text{XVI})+gI_{\theta+2}(\text{XVI})=I_\theta(\text{XV})$.
XVI	$\dfrac{dz^{\theta\eta-1}i}{\sqrt{e+fz^\eta+gz^{2\eta}}}$	−3 to +4	$\theta eI_\theta+(\theta+\frac{1}{2})fI_{\theta+1}+(\theta+1)gI_{\theta+2}=\dfrac{z^{\theta\eta}}{\eta}d(e+fz^\eta+gz^{2\eta})$. Also $e, g, \eta \to g, e, -\eta$ transforms I_θ to $I_{-\theta}$.
XVII	$\dfrac{dz^{\theta\eta-1}\sqrt{e+fz^\eta+gz^{2\eta}}\,z}{(k+lz^\eta)}$	−3 to +3	$\theta ekI_\theta+[(\theta+\frac{3}{2})fk+\theta el]I_{\theta+1}+[(\theta+3)gk+(\theta+\frac{3}{2})fl]I_{\theta+2}+(\theta+3)glI_{\theta+3}$ $=\dfrac{dz^{\theta\eta}}{\eta}(e+fz^\eta+gz^{2\eta})^{\frac{3}{2}}$. Also $eI_\theta(\text{XVIII})+fI_{\theta+1}(\text{XVIII})+gI_{\theta+2}(\text{XVIII})=I_\theta$ and $k, l, e, g, \eta \to l, k, g, e, -\eta$ transforms $I_\theta(\text{XVII})$ to $L_{-\theta}(\text{XVII})$.
XVIII	$\dfrac{dz^{\theta\eta-1}z}{(k+lz^\eta)\sqrt{e+fz^\eta+gz^{2\eta}}}$	−2 to +4	$\theta ekI_\theta+[(\theta+\frac{1}{2})fk+\theta el]I_{\theta+1}+[(\theta+l)gk+(\theta+\frac{1}{2})fl]I_{\theta+2}+(\theta+1)glI_{\theta+3}$ $=\dfrac{d}{\eta}z^{\theta\eta}\sqrt{e+fz^\eta+gz^{2\eta}}$. Also $kI_\theta(\text{XVIII})+lI_{\theta+1}(\text{XVIII})=I_\theta(\text{XVI})$ and $k, l, e, g, \eta \to l, k, g, e, -\eta$ transforms $I_\theta(\text{XVIII})$ to $I_{2-\theta}(\text{XVIII})$.

APPENDIX 4

Logometria, Part III, Problem X: The Cotes spiral, and related orbits

The necessary theory for the solution of Problem X is to be found in Newton's *Principia*, second edition (Cambridge, 1713), Book I, Section VIII. Other results from *Principia* are also used. Cotes' solution illustrates once again his powers in applying known results to obtain stylish, thorough and elegant solutions of particular problems.

Solution

S is the centre of force and the particle is projected from P, along PQ or QP with the velocity it would acquire in falling freely under the same force from a given point C, to P, along CP (see Fig. 1). Two neighbouring points on the trajectory are P, p, $P\varpi$ is perpendicular to Sp, PQ is the tangent to the trajectory at P. With centre S and any given radius SR, draw the circular arc ARr.

$$SC = c, \qquad SR = r, \qquad SP = x, \qquad p\varpi = \dot{x}, \qquad Rr = \dot{z}.$$

The velocity of the particle at P is v. The time of travel from P to p will be proportional to the area $SPp \propto (SP \cdot P\varpi) = (x^2 \dot{z})/r$; hence arc Pp will be as $(vx^2\dot{z})/r$ and since $(Pp)^2 = (P\varpi)^2 + (\varpi p)^2$, we have $[v^2 x^4 (\dot{z})^2]/r^2$ will be as $(\dot{x})^2 + [(x^2 \dot{z}^2)/r^2]$. From Propositions XXXIX and XL of Newton's *Principia*, Book I (Cotes acknowledges)

$$v^2 \text{ is as } \frac{1}{c^2} - \frac{1}{x^2}$$

and hence

$$z = \frac{er\dot{x}}{x\sqrt{x^2 - a^2 - c^2}},$$

where a is some assumed constant. (Note, put

$$\frac{\left(\dfrac{1}{c^2} - \dfrac{1}{x^2}\right) x^4 (\dot{z})^2}{r^2} = A\left(\dot{x}^2 + \frac{x^2 \dot{z}^2}{r^2}\right)$$

195

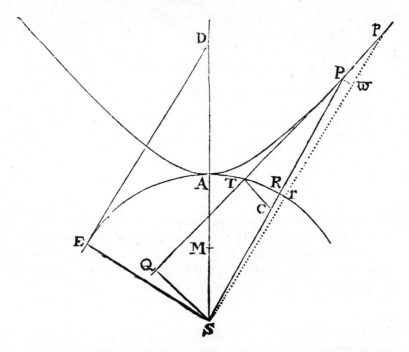

Fig. 1. From *Harmonia Mensurarum*, Cambridge University Press, 1722.

and the result follows by putting $a = c\sqrt{A}$, A being an assumed constant of proportionality.) The fluent of this quantity is obtained from Form v for $\theta = 0$, and putting $b^2 = a^2 + c^2$ the fluent is

$$z = \frac{ar}{b}(\mathrm{R}:\mathrm{T}:\mathrm{S}::b:\sqrt{x^2 - b^2}:x),$$

or by putting $r = b$, the slightly simpler form

$$z = a(\mathrm{R}:\mathrm{T}:\mathrm{S}::b:\sqrt{x^2 - b^2}:x).$$

To construct this equation: from SC drop perpendiculars SQ, CT to the tangent PQ. Triangles SRr, $ST\varpi$ are similar, therefore $\dot{z}/r\varpi = r/x$. Triangles $P\varpi p$, SQP are similar, therefore $P\varpi/x = SQ/PQ$. From these two results,

$$\frac{\dot{z}}{\dot{x}} = \frac{r \cdot SQ}{x \cdot PQ}.$$

Therefore

$$\dot{z} = \frac{SQ \cdot r\dot{x}}{x \cdot PQ} = \frac{ar\dot{x}}{\sqrt{x^2 - b^2}},$$

therefore

$$\frac{SQ}{PQ} = \frac{a}{x\sqrt{x^2-b^2}} = \left(\frac{\sqrt{b^2-c^2}}{\sqrt{x^2-b^2}}\right),$$

thus

$$\frac{b^2-SC^2}{SP^2-b^2} = \frac{SQ^2}{PQ^2}$$

and hence

$$b^2 = SQ^2 + \frac{PQ^2 \cdot SC^2}{SP^2} = SQ^2 + QT^2.$$

Thus

$$SR = r = b = \sqrt{SQ^2 + QT^2}$$

and

$$a = \sqrt{b^2-c^2}$$
$$= \sqrt{SR^2-SC^2}.$$

Finally, if z is the arc length between a fixed point A, and a variable point R,

$$z = \frac{ar}{b}(R:T:S::b:\sqrt{x^2-b^2}:x) \quad (\text{with } r=b)$$

becomes $\quad AR = \sqrt{SR^2-SC^2}(R:T:S::SR:\sqrt{ST^2-SR^2}:SP)$. Thus, the points S, P, C, T, Q being given, first the point A can be found, and hence any point on the trajectory can be constructed. The point A is easily shown to be the apse of the trajectory as follows: as R approaches A, i.e., z approaches zero, then $a(R:T:S::b:\sqrt{b^2-x^2}:x)$ approaches zero, which requires b to approach x, i.e., SP approaches SR, and the points P and R coincide in A, therefore the trajectory passes through A. Further, the particle cannot approach closer than A *atque non ulterius descendet, ne quantitas* $\sqrt{SPq - SRq}$ *fiat impossibilis.*

The time for describing the arc PA is as the area PAS.

$$\tfrac{1}{2}P\varpi \cdot SP = \frac{a\dot{x}x}{2\sqrt{x^2-b^2}}$$

and the fluent (of Form v) is intuitively seen to be $\tfrac{1}{2}a\sqrt{x^2-b^2}$ and the time of describing PA is therefore proportional to this quantity.

To compare the velocity at P with the velocity with which the particle could revolve in a circle, radius SP, centre S, under the same force reversed, velocity at P, is as $\sqrt{(x^2-c^2)/c^2x^2}$ and the required velocity for motion in a circle is that which the particle would acquire when falling freely from infinity under the reversed force, i.e., $\sqrt{1/x^2}$. Hence the two velocities are in the ratio $\sqrt{x^2-c^2}$ to c.

To construct the trajectory, PQ, S and C being given
Draw SQ, CT perpendicular to PQ. With centre S, radius ST (Fig. 1)$=\sqrt{SQ^2+QT^2}$), draw the circle ARr, making AR the arc whose secant is $\sqrt{SR^2-SC^2}$ to modulus SR, with A on the same side of PS as Q. A will be the apse of the trajectory. Points on the trajectory can now be obtained. In SA make $SM=\sqrt{SR^2-SC^2}$. Mark D any point on SA produced. Draw the tangent DE and join SE. On the other side of A mark off arc AR = angle ASE to modulus SM. Join SR and produce to P, making $SP=SD$. A point on the trajectory is P. Time from A to P is as area SAP which is as $DE \times \frac{1}{2}SM$. The velocity of the body at P will be to the velocity with which it could revolve in a circle at the same distance under an equal centripetal force, as $\sqrt{SP^2-SC^2}$ to SC.

The trajectory can also be constructed as follows
Retaining the points S, A, D as before, draw AG normal to the axis SAD. With centre S and any radius SD the circular arc DPG is described to intersect the normal AG at G. Arc DP to arc DG is taken in the ratio of SM to SA; P will be a point on the trajectory.

Notes on the foregoing construction and proof
(a)   This theorem says effectively that the space integral of the force is as the change in kinetic energy (or *vis viva*).
(b)   The constancy of areal velocity for motion under a central force is assumed, and not thought worthy of particular comment. It is established in *Principia*, Book I, Section VIII.
(c) SM is the modulus a.
(d)   This requires the division of an angle (or an arc) in any given ratio, and indicates that Cotes is concerned with practical rather than theoretical constructions.
(e)   Cotes puts this forward as a new construction, and says that by its aid he was able to construct the first, third and fifth spirals in the earlier work, the second and fourth being somewhat easier. In the Scholium to the problem under discussion (i.e., Problem x) the method is applied to demonstrate (without proof) some elegant constructions for the orbits under various initial conditions for particles moving under the inverse square law of attraction.

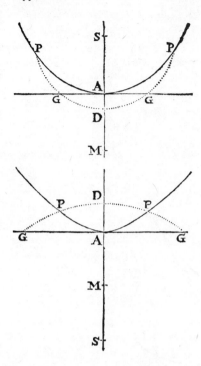

Fig. 2. From *Harmonia Mensurarum*, Cambridge University Press, 1722.

Cotes then adds four very neat constructions, as he says, from his notebook, but which he has not seen elsewhere. They are printed without proofs, but these he says are easy enough. The first and second of these two constructions were selected by Professor G. Huxley to illustrate very aptly (but without the demonstrations) the elegance of Cotes' work (see *Scripta Mathematica*, vol. 26, no. 3 (1966), pp. 231–8).

Referring to Fig. 3, the particle is in each case projected from a point *P* in a given direction *PT*, with the speed it would acquire by free fall from a given point *C* under the attraction (or in Case 4, the repulsion) of a force at *S*, under the inverse square law.

Case 1

Here *CS*, greater than *CP*, is finite, and the orbit is stated to be an ellipse.

Case 2

Here *C* recedes to infinity, and the orbit is stated to be a parabola.

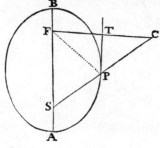

Case 1

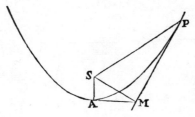

Case 2

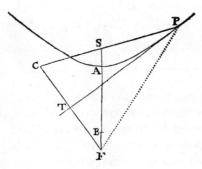

Case 3

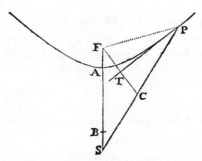

Case 4

Fig. 3. From *Harmonia Mensurarum*, Cambridge University Press, 1722.

Case 3
The initial speed is greater than in 2. Cotes says consider C to be at a greater distance than infinity, or, what is the same thing, let the particle ascend to an infinite distance from C, the force being considered as centrifugal, and then descend to P under the centripetal force. The orbit is given as the near branch of a hyperbola, (i.e., the orbit is concave to S).

Case 4
Motion under a centrifugal force: the particle is projected from P with the speed it would acquire by ascending freely from C to P. The locus is the far branch of a hyperbola.

In all cases, given that the form of the orbit is known, the constructions follow from simple well-known properties of the curves. Thus, in Cases 1, 3, and 4, the tangent bisects the focal radii, and $SP + PF$ (Case 1), $SP \sim PF$ (Cases 2 and 3) = the length of the major axis. In Case 2 the similarity of triangles ASM and MSA is a property of a parabola. What is of greater interest is the way in which Cotes' analysis, as illustrated in Problem x, would lead to information about the form of the orbit.

With the notation of Problem x, consider the fluxional equation $[(v^2 x^4 \dot{z}^2) r^2] \propto \dot{x}^2 + [(x^2 \dot{z}^2)/r^2]$, there established. This is a general equation of motion for motion under a central force, the form of the orbit being determined by v^2, which clearly depends on the initial conditions and the law of force. If the force is proportional to x^{n-1} (x being the distance from the centre S), Newton's Proposition xi (*Principia*, Book I) previously referred to, gives $v^2 \propto c^n - x^n$ and this value of v^2 substituted into the general equation above gives the differential equation of the orbit in Newtonian polar coordinates (x, z) for the various cases. Substituting then in turn, we have the fluxional equation in each case as follows, in which A is a constant of proportionality.

Cases 1, 3 and 4 are solved by reference to Form xvi, for $\theta = 0$, and Case 2 by Form v for $\theta = 0$, $\eta = 1$ in each case. It will be sufficient to show Case 1 in detail, and Case 2 briefly.

Case 1
From

$$\dot{z} = \frac{A \cdot c \dot{r} \dot{x}}{x\sqrt{-x^2 + cx - A \cdot c}},$$

Table showing fluxional equations in Cases 1, 2, 3 and 4

Case	v^2	$\dot{Z}$
1	$\dfrac{1}{x}-\dfrac{1}{c}$	$\dfrac{\sqrt{A\cdot c}\,r\dot{x}}{x\sqrt{-x^2+cx-A\cdot c}}$
2	$\dfrac{1}{x}$	$\dfrac{\sqrt{A}\,r\dot{x}}{x\sqrt{x-A}}$
3	$\dfrac{1}{x}+\dfrac{1}{c}$	$\dfrac{\sqrt{A\cdot c}\,r\dot{x}}{x\sqrt{x^2+cx-A\cdot c}}$
4	$\dfrac{1}{c}-\dfrac{1}{x}$	$\dfrac{\sqrt{A\cdot c}\,r\dot{x}}{x\sqrt{x^2-cx-A\cdot c}}$

reference to Form XVI in which

$$
\begin{array}{ccccccc}
d & e & f & g & \theta & \eta & z \\
r\sqrt{A\cdot c} & -A\cdot c & c & -1 & 0 & 1 & x
\end{array},
$$

we obtain

$$
z = r\left\{ R:T:S :: \sqrt{A\cdot c}: \frac{-A\cdot c+\frac{1}{2}cx}{\sqrt{-x^2+cx-A\cdot c}} : \frac{x\sqrt{\dfrac{c^2}{4}-A}}{\sqrt{-x^2+cx-A\cdot c}} \right\},
$$

in which I have replaced Cotes' $r:t:s$ with $R:T:S$ to avoid confusion with two r's.

Putting $z = r\theta$, we have

$$
\cos\theta = \frac{\sqrt{A\cdot c}\sqrt{-x^2+cx-A\cdot c}}{x\sqrt{\dfrac{c^2}{4}-A\cdot c}}
\tag{1}
$$

The quantity $\sqrt{-x^2+cx-A\cdot c}$ is real if

$$
\frac{c}{2}-\sqrt{\frac{c^2}{4}-A\cdot c} \leqslant x \leqslant \frac{c}{2}+\sqrt{\frac{c^2}{4}-A\cdot c}.
$$

The difference between these extremes is c and consideration of the properties of the ellipse suggest the substitutions

$$
2A = l, \qquad c = \frac{2l}{1-e^2}.
$$

Equation (1) above then yields

$$x^2(e^2 \cos^2 \theta + 1 - e^2) - 2lx + l^2 = 0, \tag{2}$$

from which $l/x = 1 \pm e \sin \theta$, the polar equation of an ellipse, θ being measured from a suitable origin.

The orbit could be plotted point by point, as in Problem x, (the Cotes spiral) but this is hardly necessary in the case of such a well-known curve as the ellipse, as Cotes' construction shows. The substitution $c = 2l/(1 - e^2)$ relates the value of c to the different conics obtainable under different initial conditions. If $e = 1$, the general equation reduces to

$$\dot{z} = \frac{\sqrt{A} r \dot{x}}{x \sqrt{x - A}}.$$

From Form v replacing

by $\quad \begin{array}{cccccc} d & e & f & \theta & \eta & z \\ r\sqrt{A} & -A & 1 & 0 & 1 & x \end{array},$

$$z = \frac{-2}{\eta e} d \mathrm{R} \left| \frac{\mathrm{R} + \mathrm{T}}{\mathrm{S}} \right. \to z = 2r\{\mathrm{R} : \mathrm{T} : \mathrm{S} :: \sqrt{A} : \sqrt{x - A} : \sqrt{x}\}$$

$$\to \cos \frac{\theta}{2} = \sqrt{\frac{A}{x}}, \qquad \text{i.e.,} \quad \frac{2A}{x} = 1 + \cos \theta,$$

a parabola, of *semi-latus rectum* $2A$, $(2A = l)$, as above.

APPENDIX 5

A manuscript in Trinity College Library, Cambridge

It is of interest to note that the date, August 11th 1703 is substantially before the date on which Dr Thomas Plume's will was witnessed, i.e., 20 October 1704. Dr D. J. Price (Annals of Science, March 28th 1952, 8, 1, page 2) suggests, reasonably enough, that Bentley was perhaps already making preparations for the establishment of an observatory.

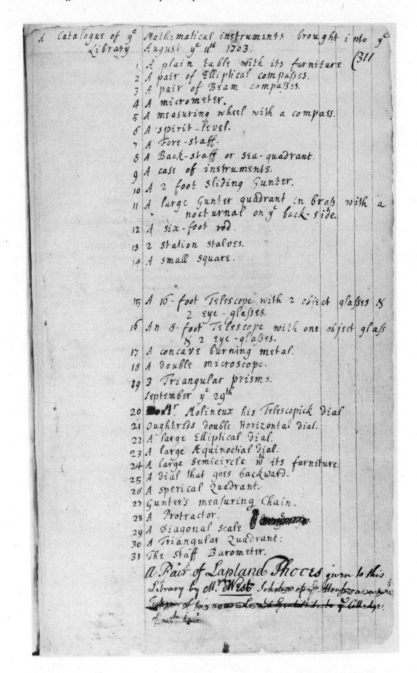

A Catalogus of ye Mathematical instruments brought into ye
Library   August ye 11th 1703.   (311

1 A plain table with its furniture
2 A pair of Elliphical compasses.
3 A pair of Beam compasses.
4 A micrometer.
5 A measuring wheel with a compass.
6 A spirit-level.
7 A fore-staff.
8 A Back-staff or sea-quadrant.
9 A case of instruments.
10 A 2 foot sliding Gunter.
11 A large Gunter quadrant in brass with a
         nocturnal on ye back-side.
12 A six-foot rod.
13 2 station stalves.
14 A small square.

15 A 16-foot Telescope with 2 object glasses &
         2 eye-glasses.
16 An 8-foot Telescope with one object glass
         & 2 eye-glasses.
17 A concave burning metal.
18 A double microscope.
19 3 Triangular prisms.
September ye 29th
20 Mr. Molineux his Telescopick dial
21 Oughtreds double Horizontal dial.
22 A large Elliphical dial.
23 A large Æquinochial dial.
24 A large semicircle wth its furniture.
25 A dial that goes backward.
26 A sperical Quadrant.
27 Gunter's measuring Chain.
28 A Protractor.
29 A diagonal scale
30 A Triangular Quadrant:
31 The staff Barometer.

A Pair of Lapland Shoes given to this
Library by Mr. West Scholar of ys House

Fig. 1. Instruments in Trinity College Library, Cambridge, 20 October 1704,
MS 1 17 12 20. (By permission of Trinity College Library, Cambridge.)

Select bibliography

[1]   Baron, M. E., *The Origins of the Infinitesimal Calculus.* London, 1969.

[2]   Cohen, I. B., *Introduction to Newton's Principia.* Cambridge, 1971.

[3]   Cotes, R., *Hydrostatical and Pneumatical Lectures,* ed. R. Smith, second edition. Cambridge, 1747.

[4]   Edleston, J., *Correspondence of Sir Isaac Newton and. Professor Cotes* (reprint of the 1850 edition). London, 1969.

[5]   Hall, R. and Tilling, L., *Correspondence of Sir Isaac Newton,* volume 5. Cambridge, 1975.

[6]   Halley, E., *Philosophical Transactions of the Royal Society,* vol. 19, no. 215. London, 1695.

[7]   Hofmann, J. E., '*Weiterbildung der Logarithmischen Reihe Mercators in England,* III', *Deutsche Mathematik,* vol. 5. Berlin, 1940.

[8]   Saunderson, N., *The Method of Fluxions.* London, 1756.

[9]   Struik, D. J., *A Source Book in Mathematics 1200 to 1800.* Harvard, 1969.

[10]   Turnbull, H. W. (ed.), *James Gregory Tercentenary Memorial Volume.* London, 1939.

[11]   Walmesley, C., *Analyse des Mesures.* Paris, 1749.

[12]   Whiteside, D. T., *The Mathematical Papers of Isaac Newton* (8 volumes). Cambridge 1967–81.

Index